Environmental Humanities: Transformation, Governance, Ethics, Law

Volume 2

Sustainability, i.e. the demand for long-term and globally practicable lifestyles and economies, is increasingly being understood as the central challenge of our time. But just as today science is often simply equated with natural science, many people think only of the natural sciences when it comes to sustainability science. Undoubtedly, natural scientific and technical knowledge of problem relationships in dealing with nature, resources and climate is important. However, technical change does not happen on its own. In addition, the ecological challenges are simply too great not to aim for a behavioural change as well as technology. This is the starting point of this series of publications. Some questions are, for example, the conditions for individual and social change, the means or governance instruments and normative (ethical and legal) issues about the ultimate goals to be pursued. Transdisciplinary approaches should play a special role, i.e. approaches that do not operate from disciplinary boundaries but from questions of content without excessive subordination to established disciplinary dogmas. It is important to the editors that the present series stands for pluralism and expressly gives room to uncomfortable, unexpected and heterodox views and methods. In times in which sustainability research in particular (also) is increasingly influenced by the interests of clients, such openness seems necessary in the interest of truly finding knowledge.

Felix Ekardt

Economic Evaluation, Cost-Benefit Analysis, Economic Ethics

A Review with Regard to Climate Change – Figures in the Sustainability Discourse

Springer

Felix Ekardt
Research Unit Sustainability and Climate Policy
Leipzig, Berlin, Sachsen, Germany

ISSN 2524-5708 ISSN 2524-5716 (electronic)
Environmental Humanities: Transformation, Governance, Ethics, Law
ISBN 978-3-030-99283-5 ISBN 978-3-030-99284-2 (eBook)
https://doi.org/10.1007/978-3-030-99284-2

This Springer imprint is published by the registered company Springer Nature Switzerland AG
The registered company address is: Gewerbestrasse 11, 6330 Cham, Switzerland

Mission Statement: New Series Environmental Humanities – Transformation, Governance, Ethics, Law

Sustainability, defined as the demand for long-term and globally practicable lifestyles and economies, is increasingly understood as the key challenge of our time. But, just as today, science is often used synonymously with natural science, and many people think only of the natural sciences when it comes to sustainability science. Undoubtedly, natural scientific and technical knowledge of problem relations is important when dealing with nature, resources and climate. However, technical change does not happen on its own. In addition, the ecological challenges are simply too big not to aim for a behavioural change as well as technology. This is the starting point of this series of publications.

Questions include, for example, the conditions for individual and social change, the tools or governance instruments, and normative (ethical and legal) issues about the ultimate goals to be pursued. Transdisciplinary approaches should play a special role, that is, approaches that do not operate based on disciplinary boundaries but based on questions of content without excessive subordination to established disciplinary dogmas. It is important to the editors that the present series stands for pluralism and expressly gives room to uncomfortable, unexpected, and heterodox views and methods. In times when particularly sustainability research is increasingly influenced by the interests of clients, this openness seems necessary in the interest of truly acquiring knowledge.

This work was first published in German in August 2018 by Metropolis Verlag.

Preface

With cost-benefit analysis, economic sciences cultivate a specific decision-making procedure, which has also been partially adopted in politics. Although economists do not experience the approach as normative, on closer examination the approach can be identified as an economic ethics. The present philosophical and at the same time transdisciplinary (with special legal and economic components) treatment examines the persuasive power of this approach using climate change as an example, as the most important sustainability issue. The objections raised against the economization of decision-making with regard to the utilitarian tradition, such as the criticism of the orientation towards weighing up options, the alleged lack of distributive justice, or the tendency to describe people in behavioral science as selfish, are hardly or not at all convincing on closer examination. In several respects, however, it turns out that cost-benefit analysis faces insoluble problems. Firstly, the theoretical basis of (hidden normative) cost-benefit analysis in philosophical empiricism does not seem tenable. This means the idea of empiricism that normative questions must be transformed into questions of factual (countable and reproduceable) preferences of people. Secondly, there are massive collisions of cost-benefit analysis with a liberal-democratic constitutional law, whose principles are universal ethical principles. This concerns both freedom rights (which must not depend on the ability of humans to pay) and the model of democracy and respect for the rule of law. Thirdly, insoluble problems of application arise for cost-benefit analyses, which are particularly (but not only) apparent in the context of climate protection, in general considerations as in the case of legislation, as well as in individual analyses, as done when constructing a coal-fired power plant. A strongly deflated cost-benefit analysis could nevertheless contribute factual material – such as partial aspects of decision consequences that can actually be depicted in monetary terms – to ethical or legal decision-making processes. In this respect, the approach appears helpful and complementary, but not beyond that.

These are the theses of the present book, which was accepted by the Faculty of Social Sciences and Philosophy of the University of Leipzig (Germany) in the summer semester 2018 as my second dissertation. The thesis is intended as a supplementary study to the analyses of the conditions of transformation, governance (i.e., the analysis of effective policy instruments for reaching given normative targets), ethics, and law of sustainability in my habilitation thesis (in a shorter English version, it has been published as Volume 1 of the present series under the title

Sustainability: Transformation, Governance, Ethics, Law). In it, my own approach to a newly formed universalistic heterodox discourse ethics, to a revised understanding of freedom, to a further developed concept of balancing, and thus to a new perspective on the possible objectivity and rationality in normative questions is dealt with in detail. It provides the basic assumption here and is only briefly touched upon in the following chapters. The present study is located at the boundary between practical philosophy, economics, and law, though the disciplinary positions are not of excessive importance according to the author's (non-Parsonsian and rather transdisciplinary) opinion. The dissertation is linked in many ways with the work and results of my Research Unit Sustainability and Climate Policy, which is based in Leipzig and Berlin, Germany, but strongly connected to the University of Rostock in many ways.

This monograph seeks to offer new perspectives. At the same time, it will contribute to new fields of study such as sustainability studies or environmental humanities with a work that offers orientation for students on critical perspectives about some core questions of sustainability in general and specifically cost-benefit analysis. In order to make this monograph usable as a textbook, it includes a glossary, an index, italic take-home messages at the end of the chapters, a box of questions for repetition at the end of each of the six main chapters, and a summary at the very end of the book summarizing all important findings.

It may be seen as an extraordinary and strange action that I have done another PhD, 18 years after my doctorate in law, 16 years after my habilitation in law (a rather transdisciplinary and sustainability-science-oriented one), and after 16 years of being a professor. It was great fun for me, as was the topic that has accompanied me since 2004. I would like to thank my esteemed colleagues Georg Meggle (University of Leipzig), Holger Rogall (HWR Berlin), and Dagmar Borchers (University of Bremen) very much for their philosophical and economic expertise and for taking on the review and engaging in this entire idea with goodwill, tolerance of content, and stimulating discussions. I would like to thank Jutta Wieding very much for her contribution to the translation of the German version, which was originally published in 2018 by Metropolis Verlag.

Leipzig, Berlin, Sachsen, Germany Felix Ekardt
August 2021

Contents

Problem and Fundaments 1

Abstract

The economic evaluation practiced in economics and beyond is an attempt to make correct individual or collective decisions derivable in a simple way, despite the immense complexity of many decision situations. Its primary form, the cost-benefit analysis, is the topic of this volume with particular regard to its chances and limits. Cost-benefit analysis is an attempt to raise the time-honored philosophical, legal and political problem of right human decision-making to a new level and to make it more precise, clearer and thus more rationally treatable. It is thus ultimately an attempt to inherit practical philosophy or ethics and to bring it into a supposedly more objective form. Cost-benefit analysis refers to the attempt to identify optimal/efficient/desirable social conditions based on people's preferences that are factually available and, if possible, translated into monetary values. This method, which is also increasingly used in politics (especially in environmental policy), takes people's factual preferences as they are. Based on this, it tries to identify social conditions with the greatest possible benefit which are translated into monetary units and the lowest possible cost. This can be understood as a sign of the highest possible individual satisfaction of needs.

To identify factual preferences and to assign a monetary value to them, cost-benefit analysis determines the willingness to pay for a certain condition, for example, for more or less climate protection. With regard to the exact determination of preferences, cost-benefit analysis acts as umbrella over a variety of methodological approaches that can be applied individually or in combination. These assess, often elicit, preferences directly (from a limited group of individuals), and hypothetically, for example, in terms of willingness to pay for a possibly

F. Ekardt, *Economic Evaluation, Cost-Benefit Analysis, Economic Ethics*, Environmental Humanities: Transformation, Governance, Ethics, Law 2, https://doi.org/10.1007/978-3-030-99284-2_1

longer (or less short) life due to increased climate action. Furthermore, it is observed whether certain advantages and disadvantages (respectively benefits and costs) of possible decisions on the market already have a market price or whether such a price can be implicitly derived from the actions of individuals.

Methodologically, this book approaches the topic on several levels. After various epistemological preliminaries and a literature analysis on the historical genesis of cost-benefit analysis and its precursors in the history of philosophy, the thesis asks whether an existing methodology (namely that of cost-benefit analysis as an implicitly practical philosophical theory) is feasible at all by going through the application of that method. Then its legal conformity is analyzed. This requires the use of the common methods of legal interpretation, i.e., the rules of interpretation of literal sense, systematics, norm purpose and history of norm formation.) Finally, the normative background assumptions of cost-benefit analysis are analytically identified and critically examined as a covert practical-philosophical theory or ethics. Essentially, ethics methodically deals with the correct argumentation, taking into account especially the basics of epistemology and logic, of course also in discussion with arguments already developed by other authors, etc. Because of their special importance, certain epistemological distinctions are explained and justified in advance, in particular the distinctions beween objective/subjective and being/should (contrary to common opinion), which are not parallel, but rather transverse to each other. Thus, for facts as well as for norms it can be asked whether they can be recognized objectively. Despite certain limitations, both can ultimately be affirmed. In addition to the epistemological analyses and the application analyses, the examination of legal conformity and the philosophical examination of the fundamentals, the work also considers certain widespread, but precisely unconvincing points of criticism of cost-benefit analysis in detail.

1.1 Problem: Economic Evaluation and the Empiricist Paradigm – Attempts to Rationalize Decisions Without Ethics

Whether it concerns morning jogging, climate protection, divorce, social policy or other very different questions, a simple-sounding but difficult to answer twofold question always arises: Which are the empirical drivers and motives for an individuals or societies (which are an aggregation of individuals) to decide for or against a specific behavior?[1] And beyond the empirical decision the question: How should

[1] On the question of the appropriate description of the motivation for action (including the question of "individual or collective") see Ekardt 2019, Ch. 2.1. Since the function of a footnote or a literature reference often remains unclear and also varies between disciplines, a comment here is appropriate. A common term such as "reference" could suggest that the statement of another person (the person quoted in the footnote) behind a statement in the text obviously also supports it is intended to prove the truth or correctness of the statement. This is, of course, only partially convincing, because then the fact that someone has already made a statement once would have to be regarded as a criterion of truth or correctness (which seems unconvincing and is also of little practical use,

one decide – and how should one deal with the conflict of differing interests contained in any decision, when practically every decision has different advantages and disadvantages in relation to different real or imaginable goals? Modern economic science, and in particular the so-called economic evaluation, provides an impressive attempt to cope with the unmanageable complexity of human decisions. This evaluation and its primary form, the so-called cost-benefit analysis, is the subject of this book in its opportunities and limitations (for more details on the terminology, see the next section). It represents an attempt to raise the long-established philosophical, legal and political problems of correct human decision-making to a new level and to make them more precise, clearer and thus more rationally treatable.[2] At the same time, it carries with it ideas of what drives people in purely factual terms. The question of what actually moves us humans and what should move us when we make decisions, describes, in a way, core areas of scientific thinking that concern various academic disciplines especially of the humanities.

The intention of economic evaluation is to depict all the advantages and disadvantages of possible decision alternatives, for example, in economic policy, nature conservation or climate policy, but ultimately in any societal issue. According to many, its goal is to do this more precisely than it has been achieved so far in the ethical, legal and political discourse. The intriguing character of this undertaking is hard to deny (many also speak in general terms of an "economization" of social life, affirmative or critical, but we will see later that such a general label hardly makes sense). Who could object to an assessment of decision options based on close examination and a recording of all their effects? In economic evaluation, the disadvantages and advantages or (in a comprehensive sense of the words) costs and benefits of various options are recorded and quantified in monetary terms as far as possible.[3] An essential assumed constraint for that is the scarcity of almost all resources.[4]

Climate change is only one example for the consideration of economic evaluation, but it is a particularly important one. For some time now, a model of civilization which was at first based on the Occident but has become increasingly globalized has been coming under scrutiny. Over the last 200 years, this model has been based

since for very contradictory statements "supporting documents" can be found). In my opinion, the usefulness of literature references is threefold: Partly it marks when someone else is the author of a thought. Sometimes it references a proof in an empirical or normative question conducted elsewhere though not presented in the present text (e.g., for reasons of space). Sometimes, however, a classification in the scientific discourse is only made possible by showing that some other voices share the expressed opinion. Which function is intended becomes usually only clear from the respective context.

[2] In this direction also Meyer 2003, p. 150; Weimann and Hoffmann 2003, p. 17 et seq.; Hoffmann et al. 2005, p. 359 et seq. On the concept of rationality see Sect. 1.4.

[3] On this, see first only briefly and concisely Baumgärtner 2003, p. 44: "According to the neoclassical theory of value, the value of a good results from equating the marginal costs of its provision with the marginal willingness to pay for the good. In this calculation, therefore, the supply or demand of a further unit of the good is considered on the basis of the currently existing quantity of the good. Thus, the quantitative measurement of the good is indispensable for determining its value."

[4] See pars pro toto Weimann and Hoffmann 2003, p. 23; Hansjürgens and Lienhoop 2015, p. 16.

largely on ahigh consumption level of fossil fuels. Fossil fuels are omnipresent in the occidental and increasingly global model of civilization – not only in petrol and electricity, but also in heating energy, in fertilizers, in almost every product, in plastics, textiles, medical products, cosmetics or in the transport of goods. As fossil fuels – and livestock production – are the main causes of global warming, biodiversity loss, disrupted nitrogen cycles, air/water/soil pollution, etc., effective climate protection will bring about significant social and especially economic changes, including the corresponding political measures. An economically guided search for optimal climate policy paths is also based, for example, on sections in the seven-yearly reports of the UN Intergovernmental Panel on Climate Change (IPCC). The IPCC reports are intended to gather the results of climate research, albeit primarily natural scientific and economic, worldwide. In the economic passages, the threat of climate change damage and the general advantages and disadvantages of possible climate policy steps, usually translated into monetary terms, are put into perspective.[5] The state to be striven for is described as optimal or efficient. A classic example of this, in addition to the more summarized IPCC reports, is the review presented to the British government by Nicholas Stern at the end of 2006, which resonated globally.[6]

Both in terms of the issues and the subject area of climate protection (or sustainability in general), this book is an in-depth examination of some important additional questions based on the author's habilitation thesis "Sustainability: Transformation, Governance, Ethics, Law."[7] At the core of that work stood the development of a relatively comprehensive anthropology, ethics, freedom/human rights theory and governance concepts and their application to questions of sustainability and climate change, taking numerous disciplines and schools into account. On the one hand, a descriptive theory of human behavior and of influencing instruments as well as a normative decision theory were developed in the humanities. On the other hand, a human-sciences theory of sustainability was established. Against this background, the present book asks the question whether economic evaluation (in general and specifically for sustainability issues) is perhaps the better normative decision theory after all. As far as possible, major repetitions of "Sustainability" are avoided in the following, which is why some of the questions that are fundamentally interesting in the rest of the field are dealt with in a rather concise manner.

The present volume is about climate and economic evaluation, however not in the sense of a broad coverage of mathematical details of various climate economic calculation models and their projections. This would lead, above all, to an examination of the highly complex scientific factual assumptions of climate and economic evaluation – since it is the global climate and decades or centuries to come that are at stake. Corresponding computer-supported models such as RICE, DICE or PAGE

[5] As an example: Nordhaus 2008, p. 5; Posner 1986, p. 85 et seq.

[6] Stern 2007, passim.

[7] Ekardt 2019, passim – there, as well as in Sect. 1.3, also in more detail on the subject area of sustainability and especially climate protection.

have already been widely discussed.[8] In contrast, the rationale of this monograph and textbook is to figure out the extent to which the economic evaluation or cost-benefit analysis as an approach can survive a critical examination or not. It is asked whether the approach can solve certain immanent problems of application, whether its (partly hidden) background assumptions are justified, and how the approach relates to the current law of liberal democracies. Further details depend on the question what exactly is meant by economic evaluation and cost-benefit analysis. In addition, these approaches will also have to be defended against some typical, but unconvincing objections (keywords for that are, e.g., individualism, homo oeconomicus, normativity, balancing, and quantification).

The fact that economic evaluation currently seems to be gaining momentum in the environmental debate, especially in the climate and nature conservation debate, cannot be overlooked. The promise that the associated demonstration of an economic value of climate and nature may possibly promote their protection is as much a part of this as the idea of decisions that are as soundly reasoned as possible and are reached with the participation of citizens and their preferences.[9] If the opportunities and limits of economic evaluations are developed and clear limits are shown, this could lead to a fundamental criticism of this approach. According to the thesis pursued here, economic evaluation will have to reestablish itself in a deflated, more modest form, in which it can then, in interaction with legally framed (and ethically founded) political decision-making programs, be justified as a supplier of certain facts. The proponents of economic evaluations have often been apostrophized as "decision-making aids," this overall assessment, however, which is anticipated here, falls far short of the comprehensive claim of the approach.[10]

The discussion about economic evaluation is ultimately part of the discourse (which has been ongoing since the classical economists and socialists of the nineteenth century) about how to approach a widely perceived process of overall social economization – or commercialization. Economization is often described *cum grano salis* as the effort to achieve a goal with the least possible effort or at the lowest possible cost; commercialization is sometimes the result of this, understood as a process in which the achievement of profit becomes increasingly the dominant goal compared to other goals (in a specific area of life).[11] The discourse about these vague and complex concepts is admittedly very complex and covers a wide variety of different aspects (affirmative or critical). In addition to the breadth of possible questions, aspects that do not necessarily belong together can also be equated. This can be observed later in detail in the relationship between economic evaluation and economic policy instruments such as taxes or cap-and-trade schemes, which are

[8] Classic, e.g., Nordhaus 2008, p. 30 et seq.; Stern 2007; on this also Nestle 2012, p. 72 et seq. and 104 et seq.; Pissarskoi 2014, p. 67 et seq.

[9] Pearce et al. 2006, p. 34 et seq.; Nestle 2012, p. 104 et seq.; Hansjürgens and Lienhoop 2015, p. 133 et seq.; Hanley and Barbier 2009, p. 307.

[10] Visible in Hansjürgens and Lienhoop 2015, p. 104 et seq.

[11] On this Kettner 2011, p. 3 et seq.; Giovanola 2011, p. 21 et seq.; Enkelmann 2011, p. 41 et seq.; Manzeschke 2011, p. 67 et seq.

often subject to fuzzy overall criticism (Sect. 2.2.8).[12] Economization can also try to describe an increasing dominance of economic thinking or other circumstances. In this respect, the present monograph and textbook would like to contribute an important aspect to any kind of reflected debate on economizing. The focus is clearly on economic evaluations or cost-benefit analyses, including their relationship to economic instruments, which have preoccupied the author for two decades – and thus on those aspects of economizing that can be relatively clearly defined.[13]

Since economic evaluations make a statement about optimal and ergo desirable conditions, a plausible initial hypothesis is that the analysis of economic evaluations is a matter of practical philosophy or ethics, even if cost-benefit analyses are mostly based in economics or, in some cases, in political science. In my terminology, ethics does not primarily refer to a specific university subject, but rather to a type of question about how individuals should behave and how societies should be structured. In the ethics practiced by philosophers, cost-benefit analysis has so far been perceived only to a limited extent, or if so, then under the heading of a criticism of utilitarianism and its tendency to weigh up everything or to balance money against life, as feared by the famous philosopher John Rawls (or possibly still in a controversy about perceived neoliberal tendencies in society, politics and science, which only partially coincides with the topic here).[14]

However, many economists would deny that their subject has anything to do with ethics at all when they examine cost-benefit calculations or the so-called efficiency of certain possible climate policy paths.[15] We will see in the following that this is not true. At the outset, representatives of cost-benefit analysis (and almost all economists) deny the possibility of answering normative questions (other than questions of fact) in an objective and rational way. And even more so, they deny that they themselves deal with such matters. However, in the further course of this study, cost-benefit analysis ultimately proves to be a normative approach, i.e., a kind of economic ethics. For the economists concerned, a "norm," as we will see more closely in the following, is usually simply what people or consumers de facto prefer. Rational are then solely quantifying (!) considerations, which bring the purely subjective and factual preferences, which themselves cannot be rationally verified, into a single currency (money) and thereby make them comparable. In other words, in this conceptual framework, one would determine the subjective value estimation of

[12] As an example: Grzeszick 2003, p. 647 et seq.; Eidenmüller 2005; with a differentiated but critical perspective Winter 2001, p. 97 et seq.

[13] See for now only Ekardt 2019, Ch. 4.

[14] Rawls 1971. It is ultimately a matter of varieties of utilitarianism in neoclassical economics as it is explicitly addressed in TEEB 2009; Eidenmüller 2005, p. 173 et seq.; Schlaudt 2016, p. 43 et seq.; Nestle 2012, p. 223 and passim. For more details, see Sects. 2.2.3 and 5.2. From an explicitly utilitarian philosophical perspective on climate change, see Broome 1991; Lumer 2000, passim. For a detailed description of modern liberalism as an intertemporally and globally just theory of liberal democracy (and for a critical assessment of its neoliberal respectively market-radical counterpart) see Ekardt 2019, Ch. 3.

[15] As an example Wink 2002; Nordhaus 2008, p. 175 et seq.; Böhringer and Welsch 2009, p. 261 et seq.

individuals for certain consequences of various decision options (also taking certain probabilities of those consequences into account).[16] This not only raises critical questions regarding some pretty obvious application problems. Rather, behind economic evaluation and cost-benefit analysis, a certain basic epistemological position could also become visible, which is obviously experienced in economic research (and far beyond) as self-evident and is therefore hardly discussed or even justified. This position can be described as epistemological empiricism, which in this case is a cipher for the idea: Only statements of facts, but not of what should be, can be rational or objective, primarily if they can be presented in a mathematically quantified and experimentally reproducible form.[17] Such a basic tendency characterizes not only today's economics, but also the great majority of other scientific disciplines. The point that this perspective is not self-evident will become apparent in the further course of this book.

It should be shown, however, that the most popular points of criticism of economic evaluations – namely the tendency of cost-benefit analysts to consistently weigh everything up and, moreover, to rely in anthropology on a homo oeconomicus (that is still popular despite all criticism) – are not convincing.[18] In this book, the following questions, which have already been briefly touched upon, are examined one after the other:

- First, the concept and methodology of economic evaluations and, in particular, of cost-benefit analysis (including their distinction from less questionable forms of economic evaluations) are considered. This includes a brief introduction to the central sustainability topic of climate protection, which is the most frequently used example in this volume. Also at the beginning of the study are comments on a number of epistemological distinctions that are important for further understanding (Chap. 1).
- Next, several (often widespread) points of criticism are analyzed (stemming from an ethical, legal, economic or civil society perspective) which – to anticipate this here – are not convincing in the end, partly because they misunderstand the cost-benefit analysis. In this context, a few remarks are made on the history of ideas behind cost-benefit analysis, especially on the early epistemological choices (Chap. 2).
- Subsequently, application problems of cost-benefit analysis such as the concrete determination of costs and benefits, the discounting question (i.e., the devaluation of processes in the future) and how to deal with uncertainties in knowledge, as is typical, e.g., for climate change, are discussed. This is the core of what is also seen in economics itself as needing further improvement and discussion

[16] These empiricist essentials can also be found, e.g., in Meggle 2010, p. 122 et seq.

[17] Aptly to this Meyer 2003, p. 150 et seq.; Rogall 2013, p. 73 et seq.; Czaniera 2004, p. 28 et seq.; Rohbeck 1978, p. 7 et seq.; Williamson 1993, p. 99 (107); Schwenker 2011, p. 38 et seq.; TAB 2015, p. 6; from a historical perspective Manzeschke 2011, p. 67 et seq.; Strunz et al. 2016; Albert 1971, p. 352 et seq.; Gorz 1989, p. 14 et seq.

[18] Exemplary for these points of criticism, e.g., Jamieson 2014, p. 115 et seq.; Moellendorf 2014, p. 90 et seq.; Klie 2010, p. 103 et seq.

regarding cost-benefit analysis. The empirical scientific data used in cost-benefit analyses, such as the idea of sustainable economic growth (Chap. 3), are also critically examined.

- In the next step, the problem of the compatibility of the cost-benefit analysis with the basic principles of a liberal-democratic legal system is analyzed, although the argumentation does not follow most typical (and rather unconvincing) points of criticism from the legal side. If one reads the liberal-democratic legal principles at the same time as a basic ethical position, one can also say something about a conflict with ethical principles (Chap. 4).
- Then, the book examines whether the basic ethical approach implied in the cost-benefit analysis can withstand a critical examination – and what role the discussions about related ethical approaches such as so-called economic ethics or varieties of utilitarianism play in this (Chap. 5).
- Finally, it is examined to what extent parts of the insights of cost-benefit analysis, possibly integrated into ethical and legal argumentation, nevertheless remain relevant, especially as a source of facts – and why it might be difficult to convince economics as a discipline of such a more reserved perspective (Chap. 6).

Briefly, the message of this volume on cost-benefit analysis is: "Not everything that can be counted counts, and not everything that counts can be counted" (Albert Einstein). Or, in other words: ethically it is possible to indicate the scope of correct decisions, but almost never exactly one decision can be described as optimal on its own, at least not in the case of real social problems as they arise today.[19] "Clear figures" from an economic point of view may suit many politicians and media representatives.[20] However, if the facts or normative ideas hidden behind the figures are wrong or dubious (which can also include unsolvable problems of application), the figures are worthless. And an objectivity suggestion that ultimately cannot be fully realized and which also collides with the foundations of liberal democratic constitutions and their ethical basis, albeit in a different way than is perhaps often perceived, takes us nowhere.[21]

Some may experience the present study as a "discourse-ethical" or "legal-philosophical" critique of cost-benefit analysis. In the opinion of the author, this is valid. However, these are not essential attributes, since, in the end, it solely matters whether or not the presently formulated criticism of the fundaments, of applicability and of legal conformity of cost-benefit analysis is true or not. Even more misleading would be the question of whether the present monograph and textbook is specifically "philosophical" or "legal" or "economic." It is true that almost all disciplines (some because of their size, others probably because of their smallness) jealously defends its limits. The question of whether someone is "a lawyer," "a philosopher," "an economist," etc. is therefore usually experienced as very important. Behind this

[19] Despite sometimes quite different turns, the critique of cost-benefit analysis in Lübbe 2015 is also centrally determined by the insight that optimal states are difficult to specify.

[20] See in detail Lang 2003, p. 96 et seq. and passim.

[21] Critical on that also Stehr and von Storch 2008, p. 19 et seq. (but only referring to uncertainties regarding facts, not norms).

is ultimately a way of thinking that Talcott Parsons paradigmatically put in a nutshell: Not the general question (ultimately detached from disciplinary boundaries) of the truth and rightness of statements, but the question of their assignability to a specific perspective of any scientific discipline is experienced as essential in this way of thinking. This line of thought is not pursued in the present volume, because the relativistic concept of descriptive truth and normative rightness contained in it is not convincing in the view held here (see Sects. 1.4 and 5.1).

1.2 Concepts: Economic Evaluation, Cost-Benefit Analysis, Related Approaches Such as Economic Ethics – and Their Methods, Especially in Climate Protection

Further considerations require, in a first step, a more detailed definition of economic evaluation and cost-benefit analysis, as well as providing some knowledge on their methodological approach. On the other hand, word formations such as economizing, financialization, monetarization, commodification, commercialization, valorization, and fiscalization are not, as has already become clear, at the centre of our interest in knowledge and are subject to a wide range of controversies. At the same time, where possible, the work will take a stand on a concretely tangible phenomenon of economizing.

The economic evaluation is, as the word explicitly states, a valuation or evaluation. Valuation in itself is an ambiguous term, even if this often remains unclear when discussing valuations and values in the sustainability debate, for example.[22] It is possible to assess class tests in mathematics, whereby one usually makes an essentially (often very simple) descriptive statement about how many arithmetical operations of the pupil were mathematically correct and how many were not.[23] It is also possible to assess scientific empirical data and thus mean that it is a matter of a more far-reaching interpretation of their meaningfulness. In this respect, too, it is ultimately a matter of a statement of facts. The meaning of the word becomes even vaguer when "evaluation" is simply used as a headline to describe one's own concluding assessments in texts of all kinds. In contrast, economic evaluations have a clearly defined subject, at least for the time being. They deal with the determination of social welfare. By social or societal welfare, one could (if one does not want to understand it merely as a synonym for an ethical generic term such as rightness or justice as a counterpart to truth in factual questions) understand differences, for

[22] Exemplified by Lang 2003, p. 96 et seq.; the entire work is about sustainability valuations, but without saying at any point what exactly is meant by valuations.

[23] We will not go into the (normative) evaluation of a legal nature here regarding school grades and, if applicable, degrees are set as prerequisites for certain professional careers and the respective grades depend on the percentage of correct answers in the examinations. Likewise, we do not address the point that subjective elements come into play in the assessment of facts ("was the arithmetic procedure in the class test also covered by the task?"). Section 1.4 discusses in more detail that the distinction objective/subjective and the distinction facts/norms are central and must be kept apart at the same time.

example, welfare as an aggregate of individual preferences, of feelings of happiness, of assessments of satisfaction as well as of ideal mental states. In economics, the latter is meant, and preferably in monetary units. We will discuss later that, in the ethical sense, this refers to evaluations or normative questions (Sect. 2.2.2). Furthermore, we will discuss what exactly is a normative statement as opposed to a statement of facts (Sect. 1.4), to what extent both are objective or subjective (Sects. 1.4 and 5.1) and what terms such as "objective" or "rational" might mean (Sect. 1.4).

According to their intention, economic evaluations of social welfare can be applied in various places.[24] Economic evaluations are relevant to abstract and general decisions and to single decisions, and they are relevant to both the public and the private sector. You can therefore evaluate a law or a general corporate strategy just as much as a single official action or a single corporate decision. In this book, the public sector's decisions that are essentially abstract and general in nature serve as an example, largely based on climate policy. However, the statements made in this book on economic evaluation are structured in a way that they also apply to single decisions of the public sector. For the private sector, on the other hand, the problems of economic evaluation developed here may not apply to the same extent due to their foundations. This will be discussed separately at the end of the study (Sect. 6.2).

Cost-benefit analysis is the main application of economic evaluation. Compared to the attempt to comprehensively determine social welfare, the complexity is significantly reduced by various specifications, in particular by the restriction to certain decision situations and by the assumption of stable (i.e., constant), context-independent and inherently consistent preferences of the participants.[25] However, the basis remains the orientation towards preferences, which in neoclassical economics (which is comprehensively dominant within the discipline) is also called subjective value theory. The cost-benefit analysis takes the factual preferences of people as they are and tries to identify social conditions measured against them with the greatest possible benefits and the lowest possible costs, understood as a sign of the highest possible individual satisfaction of needs, translated into monetary units.[26] To identify factual preferences and assign them a monetary value, the cost-benefit analysis determines willingness-to-pay for certain conditions (for the different methods see below). For example, when economists talk about death probabilities due to air pollution (and the related policies), they would in principle determine the factual preferences for this pollution, determined in very different, possibly hypothetical ways. As already mentioned, possible later changes in preferences or

[24] Hanley and Barbier 2009, p. 1 et seq.; Endres and Holm-Müller 1998, p. 15 et seq.; Brent 2017; Anderson et al. 2015.

[25] See on this Breyer and Kolmar 2014, p. 50 et seq. und 65 et seq.; Ismer 2014, p. 55 et seq.; Pissarskoi 2014, p. 60 et seq.; Hanley and Barbier 2009, p. 15 et seq.; see also Breunung 2001, p. 127 et seq.; Künzler 2012.

[26] Rogall 2013, p. 81; Hansjürgens and Lienhoop 2015, p. 63 et seq.; Hanley and Barbier 2009, p. 44 et seq.; Schlaudt 2016, p. 24 et seq.; Eidenmüller 2005, p. 326 et seq.; Anderson et al. 2015.

contradictory tendencies, such as the desire for mutually exclusive benefits, are not considered.

The determination and comparison of the costs and benefits of real or imaginable decisions is understood in the perspective of cost-benefit analysis as a particularly objective approach to decision-making questions, because money is a uniform unit of measurement and there is a clear basis for it, ostensibly without much ethical reflection on factual preferences. According to the claim, any preferences are also depicted, also immaterial values, e.g., preferences for the filter performance of soils or for the aesthetics of the landscape.[27] Furthermore, due to the complete openness of the preference approach, the possibility of retaining an option (option values), the joy about the existence of an entity such as a rare animal or plant species (existence values) or altruistic preferences in favor of other people of any kind can be considered.[28] In addition to the famous Stern Report of 2007, all this was classically carried out, for example, by a team led by Robert Costanza, who in 1997 attempted to determine the total value of all global ecosystems, considering supply services, regulatory services and cultural services provided by nature.[29]

With regard to the precise determination of preferences, the cost-benefit analysis serves as an umbrella for a variety of methodological approaches, which can be applied individually or in combination. A distinction is often made between methods of revealed preferences (i.e., implicit messages about preferences in terms of real behavior), methods of expressed preferences (both of which are benefit-based methods) and cost-based methods.[30] The perspective of those affected is taken by the so-called benefit-based methods, which, from the demand perspective, investigate preferences for better environmental quality, for example. In contrast, so-called cost-based approaches take a problem-causer or supply-oriented perspective on those who cause an environmental problem. However, cost-based approaches are also discussed on the side of those affected or users, as soon as the costs of (threatened) environmental damage are determined. And ultimately, the benefit on the supply side is another relevant parameter when it is asked what benefits an environmental problem polluter could derive if they were spared regulation, for example.

Ultimately, it seems less significant whether one speaks of "benefit-based" or "cost-based," since the benefit of one participant is usually a cost factor for other participants (and vice versa). If, for example, consideration is being given to building new coal-fired power plants or to no longer building them, different

[27] Hansjürgens and Lienhoop 2015, p. 33 et seq.; Hanley and Barbier 2009, p. 44 et seq.

[28] On these terms, e.g., TAB 2015, p. 117 et seq.; Hansjürgens and Lienhoop 2015, p. 33 et seq.; Brent 2017; see also Marggraf 2001, p. 87 et seq.

[29] Costanza et al. 1997, p. 253 et seq.; see furthermore Hansjürgens and Lienhoop 2015, p. 38 et seq. The slightly unreal character of such a broadly based undertaking (before all detailed criticism of the cost-benefit analysis in the following chapters) is pointed out at Schlaudt 2016, p. 45 et seq.

[30] The following presentation largely follows the detailed overview in Lange and Krull 2014, p. 139 and passim; Hansjürgens and Lienhoop 2015, p. 33 et seq.; UBA 2007, p. 40 et seq.; Ismer 2014, p. 55 et seq.; Jamieson 2014, p. 110 et seq.; Hanley and Barbier 2009, p. 15 et seq. and passim; Anderson et al. 2015.

stakeholders, such as residents, power plant operators, employees or the general public, will each acquire costs or benefits from the economic effects or from the consequences of climate change. This does not mean that conflict resolution is always and exclusively a zero-sum game (the history of capitalist prosperity gains over the last 250 years shows that this is hardly ever the case). Rather, it means that, if the intention of cost-benefit analysis is to be applied consistently, all conceivable advantages and disadvantages for all those affected must always be recorded. In doing so, it must be avoided to count effects twice. In this respect, the individual methods listed below are generally not all used additively, because otherwise said multiple consideration of the same aspect would occur.

With regard to the cost side for the polluters, several possible perspectives are conceivable. The opportunity cost method identifies the economic advantages of alternative uses, which must be foregone to preserve the good, as its value. The abatement-cost approach considers the costs incurred in protecting the environmental good, for example, as its value; similarly, reduction costs can also be considered. The substitution-cost method equates the costs of technical substitution of the services provided by an ecosystem (e.g., water purification) with the value of these services. The restoration-cost approach considers the costs of restoring an environmental good after its reversible destruction as its value (e.g., the costs of rewetting peatlands, which makes sense from the perspective of climate and biodiversity protection). As mentioned above, costs can also be recorded on the side of those affected or users by recording the costs of environmental damage or its avoidance or the value of an environmental good.

As far as the observed behavior of the demand side as well as the supply side is concerned, real market prices can in some cases be used. Such direct market prices arise when goods are traded on markets. However, these market prices usually do not reflect the importance of that entity for the preferences of the actors as a whole.[31] For example, abatement costs are not the same as the monetary value of an object. And the purchase price of real estate in the countryside reflects various factors, but not necessarily only the attitude towards nature. In this respect, one also speaks of the hedonic price approach or the travel-cost approach if the costs incurred for achieving a certain nature experience, for example, are meant – which then also only reflects parts of the preferences.[32]

Attempts to map resource costs overlap with what has been said, for example, when attempts are made to map the lost benefits of alternative use.[33] Since the aim is simply to steer environmental use in the most favorable economic direction, there is no real systematic separation of environmental and resource costs.[34]

---

[31] For a detailed discussion of this issue in economics, see Lange and Krull 2014, p. 138 et seq. and Hansjürgens and Lienhoop 2015, p. 50 et seq. on the various categories of benefits and costs.

[32] See on this UBA 2007, p. 89 et seq.; Hansjürgens and Lienhoop 2015, p. 52 et seq.; Fromm 1997, p. 137 et seq. und 144 et seq.; Hanley and Barbier 2009, p. 79 et seq. and 98 et seq.

[33] Discussed in detail at Lange and Krull 2014, p. 190 et seq.

[34] Aptly on this Lange and Krull 2014, p. 208.

Because of these various difficulties in working with revealed preferences (also known as morality of the markets), which become apparent during an initial description, economists sometimes use hypothetical willingness to pay for a certain good, such as a certain environmental condition or, for example, for not losing years of life (or, alternatively, the ranking of options in a non-monetary form) to determine benefits. The application problems of such and the other mentioned approaches are not yet to be discussed here (see later in Chap. 3).

According to the idea of economists, the uncertainty of future events is considered in all these approaches, and future costs and benefits are generally discounted in all of this. If preferences are not questioned, but rather observed, the question also arises in this context as to whether a certain risk aversion on the part of those involved may be assumed (see Sect. 3.2 for more details). The rate at which this could happen and the arguments with which this can be justified are very controversial, but have also been discussed extensively elsewhere and are therefore not dealt with in detail here.[35] In this respect, the analysis below is limited to the fundamental challenges of the idea of discounting (Sect. 3.3).

In all these cases, the cost-benefit analysis refers to certain predefined conditions or decision alternatives that are compared with each other. For example, it is a question of how, on a global scale, a climate policy with certain emission reductions on the one hand and unmitigated global emission of greenhouse gases on the other would be reflected in costs and benefits. The determination of costs and benefits, regardless of the method used, must be preceded by the determination of the empirical conditions of the research field, such as the relevant climate or nature conservation data. Only on this basis can costs and benefits be meaningfully investigated. After the determination of costs and benefits based on this, a sensitivity analysis follows, which attempts to determine how sensitively the overall result reacts to changes in individual benefit and cost parameters to check these again if necessary.[36]

When considering, for example, how climate change or, conversely, how an effective climate policy would translate into costs and benefits, one would have to consider a whole range of factors, some of which initially appear to be tangible in market prices, while others are not. If one considers the effects of climate change – or, conversely, of an effective climate policy – on harvests, grain prices, flood damage, the frequency of storms or the tourism industry, one is prima facie confronted, at least to a large extent, with factors where market prices exist. However, since future developments are at stake, the problem of uncertainty of future developments comes into play already. Furthermore, many other factors for which essential aspects are not available in market prices would also be relevant for determining the costs and benefits of climate change on the one hand and climate policy on the other. For example, climate change endangers many ecosystems and species, it could also affect the frequency of civil wars and wars; and the frequency of some diseases is

[35] See the analysis at Jamieson 2014, p. 115 et seq.; Stern 2007, passim; Nordhaus 2008, p. 165 et seq.; Lange and Krull 2014, p. 142; Endres and Holm-Müller 1998, p. 134 et seq.

[36] Hanley and Barbier 2009, p. 6; Hampicke 2011; Hansjürgens and Lienhoop 2015, p. 32 et seq.; Pearce et al. 2006, p. 51 et seq.

also influenced by global warming (and the continued use of fossil fuels, including associated air pollution). Conversely, the abatement-cost method could be used to try to map the disadvantages of an effective climate policy, especially for the old fossil industries, and to put them forward as costs of effective climate protection. The value of biotopes endangered by climate change could be addressed with the travel-cost method. The substitution-cost method, on the other hand, could depict the costs that would arise if ecosystem damage caused by climate change were simply accepted. In addition, the willingness of people to pay for a lower risk of losing life years through illness or war could be investigated. Ideally, at the end – whether in such an extremely large dimension or rather in relation to a single project – a statement is made as to which of the examined decision options has the best cost-benefit ratio.

The idea of optimality or efficiency conveyed by all this differs in part from concepts otherwise cultivated in the economic neoclassics[37], though different understandings are often not explicitly separated.[38] The common understanding of economic optimality is the so-called Pareto optimality. This assumes Pareto optimality as the only principle to be definitely recognized for decision situations. This is given when no further improvement is possible in a situation that puts at least one participant in a better position and no one in a worse one.[39] As a modification of this, the Kaldor-Hicks criterion is often used in economics, which also considers optimal states in which someone even though worse off could be compensated with profits of those who benefit. The comparison of advantages and disadvantages, or costs and benefits in the cost-benefit analysis is bound to the Kaldor-Hicks criterion.[40]

The cost-benefit analysis focused on in this book can be described as the main application of an economic evaluation, which is why the two are sometimes (also in my writing) latently equated. Nevertheless, it should be emphasized that some other approaches can also be described as economic evaluations in the broader sense, which are only indirectly at issue here and are not necessarily subject to the same critical questions as the cost-benefit analysis. The following approaches are meant:

- First, national accounts must be mentioned. They aim to expand the calculation of gross domestic product to include environmental issues.[41] Primarily, stocks and flows are determined in relation to environmental parameters, i.e., materially and not monetarily. Monetarization is in principle only sought if goods are available at market prices. Furthermore, environmental-economic accounting does

---

[37] See on the following Hampicke 2011; Hanley and Barbier 2009, p. 15 et seq. und 307; Weimann 1995, p. 27 et seq.; Pearce et al. 2006, p. 32; Fromm 1997, p. 14; Pissarskoi 2014.

[38] Demonstrated at von Egan-Krieger 2014, p. 100; exemplarily visible at Weimann 1995, p. 27 et seq.

[39] See, for example, Kirchgässner 2004, p. 4 et seq.; Nestle 2012, p. 78.

[40] We will not go into detail here on how market prices are exactly collected or extrapolated and which data are processed for this purpose. In particular, on the so-called marginal utility, i.e., the increase in utility by an additional unit of a good, Pindyck and Rubinfeld 2009, p. 139 et seq; Hansjürgens and Lienhoop 2015, p. 33 et seq.

[41] See in detail with further references Ekardt 2017a, p. 9 et seq.

not aim to provide recommendations for decisions like cost-benefit analysis. The same applies to calculations of the ecological effects of individual actions or lifestyles, which operate under terms such as "ecological footprint."

- The cost-effectiveness analysis or (equivalent) cost-efficiency analysis should also be mentioned. This is about gearing the economic evaluation towards a specific objective.[42] This objective is not determined by preference as in the cost-benefit analysis. Preferences are only used to determine the costs and benefits of different measures or instruments to achieve this goal, which, in itself, is fix. If necessary, this can also be limited to goods that are available at market prices. This has been discussed in detail in recent years, for example, for the implementation of the EU Water Framework Directive (which is intended to improve water quality) and in other isolated cases where such cost-effectiveness (not: efficiency) analyses are more or less suggested in legal acts.[43] In the USA, this method has been used by the government since the 1930s, especially for water-related decisions.[44]
- Differentiation questions could also arise between cost-benefit analysis and general decision theory. In economics, this term is mostly used to describe approaches of cost-benefit analysis in a more general way. In some cases, decision theory also includes the descriptive analysis of individual and collective action, which in other disciplines such as political science and sociology is also known as rational-choice research. Cost-benefit analysis, on the other hand, clearly aims to identify optimal or efficient conditions and thus to make recommendations. In various empirically oriented humanities disciplines, decision theory (beyond the ethical or legal perspective on decisions, which is not yet discussed at this point of the book) often also addresses the so-called multi-criteria analysis. In this approach, too, the aim is to find solutions to competing goals and the advantages and disadvantages of certain decision options. If this is done by weighing in a unit of measurement that is as uniform as possible, the multi-criteria analysis is more or less identical to the cost-benefit analysis. If, on the other hand, different variables are juxtaposed, it corresponds to the constitutional approach as discussed in Chap. 4 – including its ethical basis.[45]

---

[42] Discussed at Hansjürgens and Lienhoop 2015, p. 77 et seq. A certain relationship to economic evaluation can also be found in a method known as utility analysis. In this method, too, different goals and the degree of their achievement are to be coordinated optimally, but this time on the basis of the arbitrary preferences of a decisive person. With the utility analysis, this person receives assistance on how to arrange different goals, which are based on one's own preferences, in a cardinal or ordinal way and how to bring them to a good overall compromise. In contrast to the economic evaluation, however, there should be no monetarization. See for details Zangemeister 2014, p. 89 et seq.

[43] See in detail with further references Lange and Krull 2014, p. 145 et seq.; on the EU Water Framework Directive see also Ekardt and Weyland 2014, p. 12 et seq.

[44] See pars pro toto Hanley and Barbier 2009, p. 8; Lange and Krull 2014; Fehling 2004, p. 443 et seq.; Gawel 2014, p. 99 et seq.

[45] See on multi-criteria analysis Hansjürgens and Lienhoop 2015, p. 79 et seq.; TAB 2015, p. 140 et seq.

- Furthermore, in the interface of economics and philosophy, there are approaches dedicated to ethical questions of business (respectively business ethics) and which use concepts that are similar to cost-benefit analysis in that they are based on factual preferences. They are often referred to as economic ethics. For the sake of this book, economic ethics does not mean a general discussion of statements on economic issues in practical philosophy or ethics in general, which would be worth a work of its own or rather various works of their own in view of the broad and complex discourse on John Rawls, Amartya Sen or even classics such as Adam Smith and Karl Marx. With economic ethics, therefore, a quite special approach manifests (as we will see: with proximity to philosophical utilitarianism), when later its chances and limits are compared with those of cost-benefit analysis (Sect. 5.2).
- If the cost-benefit analysis is criticized, while environmental-economic accounts are not, this will also take us to a kind of overarching statement on figures in the sustainability discourse. Under this heading, further areas of questioning could of course be negotiated - and even these are not or only marginally a topic in the present volume: One could, for example, discuss the extent to which it is possible to forecast future developments in a quantified form or even just to project them or depict them in scenarios (despite the necessary under-complexity, uncertainty and the deniability of a usually extremely large number of underlying assumptions), as is more or less insinuated in the models currently widespread in economics.

Of course, cost-benefit analysis has a historical genesis. Many attribute its exact form to the neoclassical economics of a Samuelson or Walras.[46] However, the beginnings of cost-benefit analysis are found already in the nineteenth century, like in the essay "De la Mesure de l'Utilité des Travaux Publics" by the French engineer Jules Dupuit, published in 1844.[47] The fundamental considerations and ideas on which the cost-benefit analysis is based were developed even earlier; this will be dealt with in a separate section of this study (Sect. 2.1).

1.3 Further Background: Sustainability and Climate Protection

In principle, the economic evaluation can be applied to any social problem area (and, as mentioned above, also to individual and entrepreneurial actions). It is particularly strongly discussed – as in this publication – for ecological problems. In the present case, the term "sustainability issues" is used more generally. Sustainability

[46] For an example, see Becker 1976. Neoclassics can be understood here as including institutional economics, even if some neoclassicists are critical of institutional economic concepts such as transaction costs. An overarching counter-concept to neoclassics would be, for example, ecological economics: Daly 1996; Rogall 2012, p. 157 et seq. However, it will become apparent in the following that some of the points of criticism developed above also apply to them.

[47] See Dupuit 1995, p. 55 et seq.

can be understood in very different ways. In the present case, as discussed in detail elsewhere, sustainability means the overarching goal of permanently liveable and globally practicable ways of living and conducting economic activities.[48] For the purposes of this book, this general definition is sufficient. The normative question of whether sustainability is owed ethically or legally, and to what extent (and whether concepts such as strong versus weak sustainability must play a substantial role in this context), which goes beyond this mere definition, has been discussed in more detail elsewhere and appears only briefly in this volume (in Sects. 4.3 and 5.1).[49] In any case, it is uncontroversial that climate change represents a core – if not the greatest – challenge in terms of sustainability. Sustainability issues such as climate change are the example of the present study. However, the assessment of economic evaluation and cost-benefit analysis is not limited to examples of sustainability, as the problems to be discussed are of a general nature.

Building on this, here are some further basics on the example of climate change beyond what has been said, again briefly summarizing the more detailed presentation given elsewhere.[50] The dangerous drastic consequences of climate change have already been mentioned. The Paris Agreement stipulates in Art. 2 Para. 1 that global warming must be limited to well below 2, or better still 1.5 °C. For an industrialized country such as Germany with high emissions per capita that means, according to the findings of the Intergovernmental Panel on Climate Change (IPCC), that in few years a complete back out of fossil fuel use in the sectors electricity, warmth, mobility as well as with material uses (such as mineral fertilizers and plastics) has to be accomplished; also developing countries need to close up rapidly in climate protection.[51] But such goals are not in reach in Western countries such as Germany and the EU. Despite the widespread idea that the EU is a "climate pioneer" or a role model, if not an environmental pioneer overall, Germany and Europe are not: Neither in terms of absolute per-capita emissions nor in terms of the development trend (among other things because the emission reductions – at a high absolute level – are outpaced by emissions simply shifted to other countries). Against this background, the cost-benefit analysis can consider and monetarize various options and their effects, for example, climate protection in line with the Paris temperature limit mentioned above – or vice versa a strategy of continuation with the conventional economic model.

Climate goals such as the 1.5-degree limit can be achieved with certain strategies (and these strategies can then be pursued with policy instruments). Renewable

---

[48] See with further references Ekardt 2019, Ch. 1.5; see also in detail on different concepts of sustainability Rogall 2012, p. 133 et seq.

[49] See with further references Ekardt 2019, Ch. 3.

[50] See for a variety of references on the questions below in this book section and the corresponding calculations, which are not reproduced here again, Ekardt et al. 2018 and Wieding et al. 2020, based on IPCC data; see also Jackson 2016; Stengel 2011, p. 131 et seq.; Hoffmann 2015, p. 12 et seq.; Luks 2013, p. 23 et seq. und 65 et seq.; Paech 2014; Stoll-Kleemann and O'Riordan 2014, p. 34 et seq.

[51] Ibid.

resources and resource efficiency as technical perspectives alone may not be sufficient for successful sustainability and especially energy and climate change mitigation. In addition, frugality may also be needed. There is a number of reasons why it would be attractive to solve environmental problems such as climate change by purely technical means. New technology can be sold and creates jobs, whereas changes in behavior often mean taking goods out of the market and thus generally questioning the growth-oriented economic model. Moreover, a purely technical change can be more convenient and therefore easier to implement than a change in behavior. Nevertheless, there are various aspects that speak against assuming that the problem will be solved exclusively by technical means. For reasons of space, only a few are mentioned here.[52]

First, the extent of the problem must be mentioned, in this case regarding climate change. Measured against the known rates of innovation to date, it seems rather unlikely that a shift to renewable energies and energy efficiency alone could completely eliminate fossil fuels from the market in all sectors and countries within the next ten to twenty years, as the temperature limit would require. It is also unclear whether the potentials of renewable energies are always realistically imagined by their proponents. It should be noted that the extent of the problem is a global consideration, i.e., the extent to which the consumer wishes of a world society geared to economic growth and increasing prosperity can be satisfied purely technically. New discoveries of resources can only postpone this problem; in the case of climate change they even exacerbate it.

Another important factor is the expected inability of purely technical solutions to overcome greenhouse gas emissions, for example, in food supply. This brings us to perhaps the most important point: In the long term, environmental problems other than climate change must be solved to ensure the existence (as well as economic and world peace) of the country. Those other environmental problems, however, are often far worse off in terms of technical solutions than climate change. Central examples are the damaged ecosystems with loss of biodiversity, disturbed nitrogen cycles and soil degradation. Solutions foremostly include that people need to withdraw from land and curb agricultural production. This implies an end to ever larger living spaces and steadily increasing consumption rates for livestock products; likewise, lower yields per hectare due to the elimination of mineral fertilizers must not be compensated for by ever increasing land use. Also, it will hardly be possible to convert the material basis of all convenience goods completely to renewable resources, inter alia because they are not as inexhaustible as one could think.

Thus, despite all uncertainties about future technical developments, there is much to be said for the fact that technology alone will not be sufficient to meet the ecological challenges if we do not want to run into existing problems. Because of the magnitude of the challenge, this is by no means merely a distributive problem; it is not sufficient to begin with if only a few very wealthy people limiting themselves, as the figures above should make sufficiently clear. It is beyond the scope of

[52] On all this see again the references from the last footnotes, especially Stoll-Kleemann and O'Riordan 2014, p. 34 et seq. and Ekardt 2019.

this book to show that it is not possible to escape all this by relying on hypothetical wonder technologies such as huge afforestations to bind climate gases; in short, these do not achieve the other environmental goals, but rather intensify them, and their effectiveness as a climate protection measure is uncertain in the short term.

1.4 Epistemological and Methodological Remarks: Objective vs. Subjective, Is vs. Ought, Genesis vs. Validity, Normative vs. Instrumental and Theoretical Rationality

Some epistemological preliminary remarks for further investigation are appropriate. These are usually not found in investigations to the cost use analysis. However, in the further process it will be repeatedly necessary to refer to those foundations. While cost-benefit analysis is an attempt to take decisions on an objective basis, we first need to ask what can actually be considered as objective. Therefore, at least brief remarks on the distinction objective/subjective as well as on is/ought and on the distinction genesis/validity are important.[53]

The fundamental existence of objective facts and the fundamental existence of a difference between is statements and ought statements, since they are not addressed separately in the debate on cost-benefit analyses, are assumed to be given in the present case after they have been presented in more detail elsewhere.[54] It does not follow logically from observations of nature on imminent damage that these observations are to be welcomed or criticized; even this basic factor is often not sufficiently present in the economic and scientific debate.[55] The distinction between is and ought is the basis for the existing impossibility to infer norms directly from facts (often this is then also called "naturalistic fallacy"): From climate change (fact), for example, does not follow its commandment or prohibition (norm), even if many public discussions take place in this way. Furthermore, it is also not possible to draw a conclusion about an "is" based on an "ought": The idea that something is desirable does not prove that it is empirically the case.[56] To make evaluations, the

---

[53] The account given here follows, abbreviated and occasionally supplemented, Ekardt 2019, Ch. 1.6.

[54] See on this, with an account of various literary positions, Ekardt 2019, Ch. 1.6 (including a variety of references to voices that do not precisely make these distinctions); similarly, here, Nagel 1979.

[55] Though subliminally suggested, always recurring in the background in Stern 2009, p. 86 et seq. Meanwhile, Schlaudt 2014, p. 45 et seq. attempts a pragmatic new approach to the topic. However, the distinctions genesis/validity and facts/norms are paralleled in contrast to the present text. Schlaudt is only concerned with the former distinction.

[56] See, for example, Caspari and Schefold 2011, p. 277 et seq. The criticism in von Kutschera 1982, p. 29 et seq. that is-ought fallacies are not always to be qualified as errors is not convincing. First, the idea that normative statements contain empirical elements like probability statements is skewed, because norm and subsumption material are mixed here. The second indication given there, "ultra nemo posse obligatur" and that impossibility is an empirical category from which

existence of a fact is not enough – you need an evaluation criterion, i.e., a norm that says you should not kill people or you should preserve the human basis of life and therefore a stable global climate. It is true that facts indicate the area of application or the subsumption material of an evaluation. An example: The sentence "People have a basic right to the ecological subsistence level" is a sentence about what should be done or a sentence of ethics or law. Likewise, the concretization that "that right always prevails in the balancing when it collides with the basic right to entrepreneurial freedom of the people living today" or the further concretization "entrepreneurial freedom contains first of all the claim to be allowed to operate coal-fired power plants, which is then, however, subject to the weighing with colliding rights."

To go into the distinction between objective and subjective, one must first define the terms: "objective" here means "non-subjective," i.e., independent from certain perspectives, cultural backgrounds or attitudes – i.e., universally and everywhere valid (although in the present discussion it will not be reopened whether the more appropriate term would be intersubjective instead of objective). The differentiation "objective" versus "subjective," contrary to an exercise that can be encountered time and again, is by no means per se synonymous with the distinction "is" versus "ought." Rather, the question arises both for statements of fact and for statements of value (or ought), whether they can be objective or always only subjective. It would be too simple to say that a statement of fact is "science" and a statement of valuation "personal opinion." Rather, science versus opinion is represented by the distinction between objective and subjective – and this distinction can be made in relation to facts as well as to norms.

In this context, "objective norms" does not mean that a standard is supported by anyone on a purely factual basis; the latter would then be the respective subjective standard of the persons concerned. In turn a fact that does not necessarily have to prove that the standard is correct in the objective sense (example: Hitler subjectively followed the norm of killing all Jews, but this does not say anything about the objective validity of that norm). Furthermore, when one brings a personal viewpoint into disputes about facts, one makes a subjective, but not automatically a normative statement, because the statement remains one about facts and not about whether something is desirable or not.[57]

Therefore, objectivity does not mean that the factual emergence of norms can be empirically explained (for example, by analyzing one's economic interests, one's genes, or one's difficult childhood) – or that it can be established that someone cultivates certain normative views. Nor does it mean that the factual elements of a normative doctrine – such as the existence of climate change within the framework of a normative theory of climate justice – can be objectively verified. Rather, it is a question of the objectivity of the normative statements themselves. For example, it

something normative can be derived, is in my opinion also based on a misunderstanding, because in this respect it is a matter of a clearly justified exception, if one does not want to lead the meaning of norms as a whole ad absurdum; on this, Ekardt 2019, Ch. 3.1.

[57] See as an example for this misunderstanding Weimann and Hoffmann 2003, p. 23 et seq.; Baumgärtner 2003, p. 43 et seq.

is not about the factual historical genesis of human rights (in which, for example, the economic interests of the emerging middle classes played a certain role), but rather about the normative justification why it is right to have human rights in a basic social order. Of course, facts can also be twisted (e.g., by denying climate change or exaggerating the severity of its consequences), because one hopes to promote certain values (e.g., freedom of property or, vice versa, an end to capitalism), and of course there is always this appropriation of facts for normative goals, but the distinction remains valid.[58]

Whether there are objectively valid norms and facts has, according to what has been said, nothing to do with the (correct) observation that, purely factual, our subjective views repeatedly get in the way of our knowledge of facts and norms. Of course, people have a tendency to take a subjective perspective. However, this does not prove that objectivity – for example, through careful examination and discourse with others – is absolutely impossible.[59] To give an example from the realm of facts: There may be natural scientists who speak out for or against the existence of human-made climate change because they expect financial benefits. This would distort their statements not objectively, but subjectively. However, this does not mean that it is not possible to gain objective and undistorted insights into climate change. The identification of factually often distorted perspectives presupposes that there are objective perspectives. Otherwise, the distortion of the subjective perspectives could not be reasonably determined.

As far as statements of facts can be objectively true and (nothing has been said on this here so far) statements of norms can be objectively correct, objective means that they are rationally recognizable and therefore everybody could at least see them. For this reason, the idea of objectivity is directly intertwined with the idea of reason as a mode of acquiring knowledge that is fundamentally accessible to all human beings (this becomes clearer in Sect. 5.1, when the possibility of objective norms is more closely analyzed). *Reason or rationality in this case means the ability to make decisions on questions on the basis of reasons.* Where it concerns the validity of ethical or legal principles of justice, we speak here of "normative reason." In contrast, instrumental and theoretical reason derive from facts. Instrumental reason deals with which means most effectively implement a norm that is assumed to be correct, such as a certain climate goal (or even a completely selfish goal such as theft) – for example, through emissions trading. Theoretical reasoning deals with fact-finding without concrete reference to action, as is the case in scientific climate research, for example. Facts of theoretical reason are included in decisions of normative reason as subsumption material.[60]

---

[58] See Ekardt 2021, § 1 D. II. also on (among others in discussion with Scarano 2012 and von der Pfordten 1993) why it does not promise any findings to distinguish once again the category norm or ought from a category values.

[59] Likewise, for exactly this differentiation the (in this regard often misunderstood) classic Berger and Luckmann 2011.

[60] In more detail and with further references Ekardt 2019, Ch. 3.1 and Ekardt 2021, §§ 1 D. II., 3 F.-G. (also on the classical question of the so-called theories of truth, i.e., the criterion for a factual

The question of whether it is at all desirable to avert the consequences of climate change, such as the threat of resource wars, millions of deaths, high economic damage, etc., is thus a question for normative reasoning – including the question of which distributional standards are to be applied, how historic emissions are to be dealt with, etc. In contrast, mere strategic recommendations or instruments have no normative character. Example: The statement "taxes on fossil fuels should be increased" is normative only with regard to the goal contained therein, i.e., the reduction of greenhouse gases for the sake of human rights, whereas the statement that fuel price increases are an effective means of achieving the goal is fact-based.

Economists, however, usually only accept the balancing component of normative rationality; for economists, the object of balancing tends to be preferences expressed in monetary values.[61] The balancing is then a quantifying and empirical process, which ultimately offsets the willingness to pay for different decision options against each other. It will become apparent in the further course of this book that narrowing of the concept of rationality in this way is not convincing, since it is a central aspect of the discussion on cost-benefit analysis.

Based on the distinction between is and ought, theories of how societies and people should be shaped must also be separated from those theories of behavior or society that merely explain human behavior (normative justification versus descriptive explanation). Normative theory is usually referred to as ethics or theory of justice.[62] More generally one can speak of practical philosophy. In the present case, justice means, by definition, the rightness of a social order. Justice is therefore not discussed here as something additional, which can be formulated as an additional criterion after issues like wealth, even if it often sounds that way in economic discourses.[63] Moreover, it does not mean mere social distributive justice. Any idea of how a society should be (and may it simply be: society should be as rich as possible, whereby we express everything in money, and the distribution of wealth is irrelevant) is per se a concept of justice, whether right or wrong. Concepts of a successful society treat justice per se in this definitional sense, just as physics or biology per se deal with descriptive truth (even if single research results may not be true, but rather untrue, and thus miss the point). It will be discussed later whether cost-benefit anal-

statement to be considered objective, among others in discussion with Rorty 1989 and Scherzberg 1994, p. 25 et seq.). On the partly very contradictory conceptual understandings of rationality and forms of rationality – in addition to the following remarks on the conceptions within cost-benefit analysis – also Hahn 2013, p. 55 et seq.

[61] See, for example, Simon 1955 and Simon 1993 (who even doubts the rationality of balancing). The restriction to subjective values is also found in Meggle 2010, p. 122 et seq. Often, for instance in Steinberg 2013, p. 101, the economic variant of rationality is also called "purpose rationality" following Max Weber, whereby this term is supposed to include quantified balancing and the effective choice of means.

[62] Whereby this can contain instructions for action on a social as well as on an individual level, and, e.g., a central theme is often the examination or justification of a certain (for instance the liberal-democratic) legal order. Ethics is therefore not only a maxim of action in the private sphere; apparently differently understood in Rogall 2012, p. 185 et seq.

[63] As an example Panther 2006, p. 21 et seq.; Nutzinger 2006.

ysis really remains an empirical approach in its basic construction in addition to the various facts that it indisputably attempts to ascertain, or whether it is rather a normative approach, i.e., whether it deals with justice (Sect. 2.2.2).[64]

The fact that factual statements, for example, on anthropology or climate data, may be in part uncertain and difficult to prove, but can be fundamentally true and thus objectively justified, i.e., rational, is not disputed by cost-benefit analysts, as far as can be seen. It is less clear, however, whether ethical and legal norms can also be correct and objective/rational. As already stated in the introduction, many economists explicitly or implicitly assume that only economics and natural sciences, or at least only empirical sciences, can be rational. Whether this noncognitivism can withstand a critical examination will therefore be examined later (Sects. 4.2 and 5.1).

The methodological approach used in the present study is derived indirectly from what has been said. Economists typically ask which "methods" are used. However, even philosophical approaches that are based on empirical facts (such as preferences, happiness, cultural contexts and traditions, lived practices, etc.) do not use an explicit methodology to collect those empirical data. Utilitarians, for example, do not generally conduct empirical happiness research, and contextualists are no sociologists or historians who empirically survey the norms that actually prevail as cultural context in a society. This is a difficult problem to cure if one wants to include empirical facts in ethics in a substantiating manner, because even if this approach were correct, it would be worthless if the required facts were not determined correctly. With regard to cost-benefit analysis, although a clear methodology is available, it turns out in the further course (Chap. 3) that it leads to insoluble application problems. At the same time, it becomes clear what is being done in this case. It is methodically examined whether an existing methodology (namely that of cost-benefit analysis) is feasible at all by playing through the application of that method. Furthermore, its legal conformity is analyzed. And the normative background assumptions of cost-benefit analysis are identified and critically examined. Ultimately, ethics methodically deals with correct reasoning, taking into account the fundamentals of epistemology and logic, and of course also considering arguments already developed by other authors etc.; the further illustrative material of the text will help to clarify this. The legal methodology, which plays a role when the relation between cost-benefit analysis and law is discussed, will be described separately later (Sect. 4.1). It shows the content of the legal norms against which cost-benefit analysis will be measured: namely, the basic norms of liberal democracy. As far as the historical foundations of cost-benefit analysis are finally uncovered, the present study concentrates on a literature analysis.

[64] In contrast, approaches to justification that are (partly) similar to the own approach pursued in the present book, without dealing directly with cost-benefit analysis, develop Alexy 1995, p. 127 et seq.; Illies 2003, p. 129 et seq.; Kuhlmann 1985; partly Habermas 1983, p. 56 et seq.; implicitly also Ott and Döring 2004, p. 91 et seq.

Repetition Questions

1. What do the terms economic evaluation and cost-benefit analysis mean, and how do they relate to each other and to similar terms? (Sects. 1.1 and 1.2)
2. What exactly does the term sustainability mean? (Sect. 1.3)
3. What is meant by "empiricism" in epistemology? (Sect. 1.1)
4. What are the major drivers of various ecological challenges? (Sect. 1.3)
5. What are the three fundamental sustainability strategies, how do they differ, and for what reasons could purely technological strategies reach their limits? (Sect. 1.3)
6. Which are the basic epistemological distinctions? (Sect. 1.4)
7. Why can the possibility of objective truth not be denied in epistemology, and what different forms of rationality or reason exist? (Sect. 1.4)

References[65]

Albert H (1971) Modellplatonismus. Der neoklassische Stil des ökonomischen Denkens in kritischer Betrachtung. In: Topitsch E (ed) Logik der Sozialwissenschaften, 7th edn. Königstein, p 352 et seq

Alexy R (1995) Recht, Vernunft, Diskurs. Frankfurt a.M

Anderson M, Teisl M, Noblet C, Klein S (2015) The incompatibility of benefit–cost analysis with sustainability science. Sustain Sci 10(1):33–41. https://doi.org/10.1007/s11625-014-0266-4

Baumgärtner S (2003) Warum Messung und Bewertung biologischer Vielfalt nicht unabhängig voneinander möglich ist. In: Weimann J, Hoffmann A, Hoffmann S (eds) Messung und ökonomische Bewertung von Biodiversität – Mission impossible? Metropolis, Marburg, p 43 et seq

Becker G (1976) The economic approach to human behavior. University of Chicago Press, Chicago

Berger P, Luckmann T (2011) The social construction of reality: a treatise in the sociology of knowledge. Reprint New York

Böhringer C, Welsch H (2009) Effektivität, Fairness und Effizienz in der internationalen Klimapolitik: Contraction and Convergence mit handelbaren Emissionsrechten. Jahrbuch Ökologische Ökonomik:261 et seq

Brent R (2017) Advanced introduction into cost-benefit analysis. Cheltenham, New York

Breunung L (2001) Umweltschutz unter dem Optimalitäts- und Effizienzpostulat. Die wohlfahrtsökonomische Konzeption in soziologischer Sicht. In: Gawel E (ed) Effizienz im Umweltrecht. Grundsatzfragen einer wirtschaftlichen Umweltnutzung aus rechts-, wirtschafts- und sozialwissenschaftlicher Sicht. Nomos, Baden-Baden, p 127 et seq

Breyer F, Kolmar M (2014) Grundlagen der Wirtschaftspolitik, 4th edn. Mohr Siebeck, Tübingen

Broome J (1991) Counting the cost of global warming. Cambridge University Press, Cambridge

Caspari V, Schefold B (eds) (2011) Wohin steuert die ökonomische Wissenschaft? Ein Methodenstreit in der Volkswirtschaftslehre. Campus Verlag, Frankfurt a.M

Costanza R et al (1997) The value of the world's ecosystem services and natural capital. Nature:253 et seq

[65] In accordance with legal practice, parliamentary, governmental and EU Commission documents as well as laws and judgments are not listed in the bibliography, as they can be found unequivocally based on the reference given in the continuous text or via the general search engines. The last access date for all Internet sources is 31.10.2021.

Czaniera U (2004) Vernünftige Normen statt moralischer Fakten. In: Lütge C, Vollmer G (eds) Fakten statt Normen? Zur Rolle einzelwissenschaftlicher Argumente in einer naturalistischen Ethik. Nomos, Baden-Baden, p 28 et seq

Daly H (1996) Beyond growth. The economics of sustainable development. Beacon Press, Boston

Dupuit J (1995) De la mesure de l'utilité des travaux publics (ursprünglich 1844). Revue française d'économie:55 et seq

Eidenmüller H (2005) Effizienz als Rechtsprinzip. Möglichkeiten und Grenzen der ökonomischen Analyse des Rechts, 3rd edn. Mohr Siebeck, Tübingen

Ekardt F (2017a) Natural capital accounting and the mobilization of financial resources for biodiversity, Analyses for the German Federal Government. Leipzig. http://felix-ekardt.eu/files/texts/Biodiv-Evaluation-Instruments.pdf

Ekardt F (2017b) Wir können uns ändern. Gesellschaftlicher Wandel jenseits von Kapitalismuskritik und Revolution. München

Ekardt F (2019) Sustainability. Transformation, governance, ethics, law. Springer, Dordrecht

Ekardt F (2021) Theorie der Nachhaltigkeit. Ethische, rechtliche, politische und transformative Zugänge – am Beispiel von Klimawandel, Ressourcenknappheit und Welthandel, 4th edn. (= 3rd ed. of the new edition). Baden-Baden

Ekardt F, Weyland R (2014) Neues vom wasserrechtlichen Verschlechterungsverbot und Verbesserungsgebot: Rechtsinterpretations- und Rechtswirkungsfragen im Spiegel neuerer Urteile. Natur und Recht 36:12 et seq

Ekardt F, Wieding J, Zorn A (2018) Paris agreement, precautionary principle and human rights: zero emissions in two decades? Sustainability 10:2812

Endres A, Holm-Müller K (1998) Die Bewertung von Umweltschäden. Theorie und Praxis sozioökonomischer Verfahren. Kohlhammer, Stuttgart et al

Enkelmann WD (2011) Zwischen Ökonomie, Kommerzialität und Idealismus. Das zoon logon echon – Aristoteles' Konzeption des homo oeconomicus. In: Kettner M, Koslowski P (eds) Ökonomisierung und Kommerzialisierung der Gesellschaft. Wirtschaftsphilosophische Unterscheidungen. Wilhelm Fink, Munich, p 41 et seq

Fehling M (2004) Kosten-Nutzen-Analysen als Maßstab für Verwaltungsentscheidungen. Verwaltungsarchiv:443 et seq

Fromm O (1997) Möglichkeiten und Grenzen einer ökonomischen Bewertung des Ökosystems Boden. P. Lang, Frankfurt a.M

Gawel E (2014) Konzept zur „Berücksichtigung" von Umwelt- und Ressourcenkosten nach Art. 9 WRRL, Verwaltungsarchiv:99 et seq

Giovanola B (2011) Ökonomisierung und Kommerzialisierung als Problem der Philosophie. In: Kettner M, Koslowski P (eds) Ökonomisierung und Kommerzialisierung der Gesellschaft. Wirtschaftsphilosophische Unterscheidungen. Wilhelm Fink, Munich, p 21 et seq

Gorz A (1989) Kritik der ökonomischen Vernunft. Sinnfragen am Ende der Arbeitsgesellschaft. Rotbuch, Berlin

Grzeszick B (2003) Lässt sich eine Verfassung kalkulieren? Juristenzeitung 58:647 et seq

Habermas J (1983) Moralbewusstsein und kommunikatives Handeln. Frankfurt a.M

Hahn S (2013) Rationalität. Eine Kartierung. Münster

Hampicke U (2011) Climate change economics and discounted utilitarianism. Ecol Econ 72:45 et seq

Hanley N, Barbier E (2009) Pricing nature. Cost-benefit analysis and environmental policy. Elgar, Cheltenham

Hansjürgens B, Lienhoop N (2015) Was uns die Natur wert ist. Potenziale ökonomischer Bewertung. Metropolis-Verlag, Marburg

Hoffmann U (2015) Can green growth really work – and what are the true (socio-)economics of climate change? UNCTAD, Berlin

Hoffmann A, Hoffmann S, Weimann J (2005) Irrfahrt Biodiversität. Eine kritische Sicht auf europäische Biodiversitätspolitik. Metropolis-Verl, Marburg

Illies C (2003) The grounds of ethical judgement – new transcendental arguments in moral philosophy. Clarendon Press, Oxford

Ismer R (2014) Klimaschutz als Rechtsproblem. Steuerung durch Preisinstrumente vor dem Hintergrund einer parallelen Evolution von Klimaschutzregimes verschiedener Staaten. Mohr Siebeck, Tübingen

Jackson T (2016) Prosperity without growth, 2nd edn. London

Jamieson D (2014) Reason in a dark time. Why the struggle against climate change failed – and what it means for our future. Oxford University Press, Oxford

Kettner M (2011) Ein Vorschlag zur Unterscheidung von Ökonomisierung und Kommerzialisierung. In: Kettner M, Koslowski P (eds) Ökonomisierung und Kommerzialisierung der Gesellschaft. Wirtschaftsphilosophische Unterscheidungen. Wilhelm Fink, Munich, p 3 et seq

Kirchgässner G (2004) Wertfreiheit und Objektivität in den Wirtschaftswissenschaften – Mythos oder Realität? Discussion Paper No. 2004-15. St. Gallen

Klie A (2010) Die Bewertung von Umweltgütern mittels Zahlungsbereitschaft. Woran Kosten-Nutzen-Analysen scheitern. GAIA:103 et seq

Kuhlmann W (1985) Reflexive Letztbegründung. Freiburg et al

Künzler A (2012) The jurisprudence of welfare maximization. Embracing complexity and uncertainty in the environmental, health and safety context. Archiv für Rechts- und Sozialphilosophie 98:1 et seq

Lang A (2003) Ist Nachhaltigkeit messbar? Eine Gegenüberstellung von Indikatoren und Kriterien zur Bewertung nachhaltiger Entwicklung unter Berücksichtigung der Rahmenbedingungen in Deutschland und Frankreich. Ibidem Verlag, Stuttgart

Lange C, Krull D (2014) Infrastrukturelle Wasserdienstleistungen. Zur Funktion kostendeckender Wasserpreise nach Art. 9 der EG-Wasserrahmenrichtlinie und ihre mögliche Anwendung auf den Aufstau von Gewässern als infrastrukturelle Wasserdienstleistung für Binnenschifffahrt und Wasserkraft. Metropolis-Verl, Marburg

Lübbe W (2015) Nonaggregationismus. Grundlagen einer Allokationsethik. Münster

Luks F (2013) Die Zukunft des Wachstums. Theoriegeschichte, Nachhaltigkeit und die Perspektiven einer neuen Wirtschaft, 2nd edn. Metropolis-Verl, Marburg

Lumer C (2000) The greenhouse: a welfare assessment and some morals. University Press of America, Lanham

Manzeschke A (2011) „Ökonomisierung" – Klärungsbedürftigkeit und Erklärungskraft eines Begriffs. In: Kettner M, Koslowski P (eds) Ökonomisierung und Kommerzialisierung der Gesellschaft. Wirtschaftsphilosophische Unterscheidungen. Wilhelm Fink, Munich, p 67 et seq

Marggraf R (2001): Nutzungsunabhängiger Wert der Natur und ökonomisches Verhaltensmodell. Jahrbuch Ökologische Ökonomik:87 et seq

Meggle G (2010) Handlungstheoretische Semantik. Berlin et al, De Gruyter

Meyer M (2003) Überlegungen zur Rationalität institutionenökonomischer Modelle. In: van Aaken A, Schmidt-Lübbert S (eds) Beiträge zur ökonomischen Theorie im Öffentlichen Recht. Deutscher Universitätsverlag, Wiesbaden, p 149 et seq

Moellendorf D (2014) The moral challenge of dangerous climate change. Values, poverty, and policy. Cambridge University Press, Cambridge, MA

Nagel T (1979) The limits of objectivity. Harvard University Press, Cambridge, MA

Nestle I (2012) The costs of climate change in the agricultural sector. A comparison of two calculation approaches, Dissertation, Flensburg

Nordhaus W (2008) A question of balance. Weighing the options on global warming policies. Yale University Press, New Haven et al

Nutzinger H (ed) (2006) Gerechtigkeit in der Wirtschaft – Quadratur des Kreises? Metropolis-Verl, Marburg

Ott K, Döring R (2004) Theorie und Praxis starker Nachhaltigkeit. Metropolis-Verl, Marburg

Paech N (2014) Liberation from excess. Munich

Panther S (2006) Gerechtigkeit in der Wirtschaft. In: Nutzinger H (ed) Gerechtigkeit in der Wirtschaft – Quadratur des Kreises? Metropolis-Verl, Marburg, p 21 et seq

Pearce D, Atkinson G, Mourato S (2006) Cost-benefit analysis and the environment. Recent development. Paris

Pindyck R, Rubinfeld D (2009) Mikroökonomie, 7th edn. Munich
Pissarskoi E (2014) Gesellschaftliche Wohlfahrt und Klimawandel. Umgang mit normativen Annahmen und Ungewissheiten bei der klimaökonomischen Politikberatung. Oekom-Verl, Munich
Posner R (1986) Wealth maximization revisited. Notre Dame J Law Ethics Public Policy:85 et seq
Rawls J (1971) A theory of justice. Cambridge University Press, Cambridge, MA
Rogall H (2012) Nachhaltige Ökonomie. Ökonomische Theorie und Praxis einer Nachhaltigen Entwicklung, 2nd edn. Marburg (3rd ed. angekündigt für 2018)
Rogall H (2013) Volkswirtschaftslehre für Sozialwissenschaftler. Einführung in eine zukunftsfähige Wirtschaftslehre, 2nd edn. Springer, Wiesbaden
Rohbeck J (1978) Egoismus und Sympathie. David Humes Gesellschafts- und Erkenntnistheorie. Campus-Verlag, Frankfurt a.M. et al
Rorty R (1989) Contingency, irony, and solidarity. Cambridge University Press, Cambridge, MA
Scarano N (2012) Wozu Metaethik? FIPH-J 19:1 et seq
Scherzberg A (1994) Die Öffentlichkeit der Verwaltung. Nomos, Baden-Baden
Schlaudt O (2014) Was ist empirische Wahrheit? Pragmatische Wahrheitstheorie zwischen Kritizismus und Naturalismus. V. Klostermann GmbH, Frankfurt a.M
Schlaudt O (2016) Wirtschaft im Kontext. Eine Einführung in die Philosophie der Wirtschaftswissenschaften in Zeiten des Umbruchs. Vittorio Klostermann GmbH, Frankfurt a.M
Schwenker B (2011) Ein Plädoyer für (mehr) Wachstum. In: Roland Berger Strategy Consultants (ed) Der Preis des Wachstums. Nachhaltige Entwicklung versus BIP-Orientierung. Wien, p 38 et seq
Simon H (1955) A behavioral model of rational choice. Q J Econ:99 et seq
Simon H (1993) Homo rationalis. Frankfurt a.M
Stehr N, von Storch H (2008) Anpassung und Vermeidung oder von der Illusion der Differenz. Gaia 17:19 et seq
Steinberg R (2013) Die Repräsentation des Volkes. Menschenbild und demokratisches Regierungssystem. Nomos, Baden-Baden
Stengel O (2011) Suffizienz. Die Konsumgesellschaft in der ökologischen Krise. oekom verlag, Munich
Stern N (2007) Stern review. The economics of climate change. HM Treasury, Cambridge
Stern N (2009) A blueprint for a safer planet. How to manage climate change and create a new era of progress and prosperity. Bodley Head, Cambridge
Stoll-Kleemann S, O'Riordan T (2014) The sustainability challenges. Environment 3:34 et seq
Strunz S, Klauer B, Ring I, Schiller J (2016) Between Scylla and Charybdis? On the place of economic methods in sustainability science. Sustain Sci. https://doi.org/10.1007/s11625-016-0407-z
TAB (Büro für Technikfolgenabschätzung beim Deutschen Bundestag) (2015) Inwertsetzung von Biodiversität. Wissenschaftliche Grundlagen und politische Perspektiven. Berlin
TEEB (2009) The economics of ecosystems and biodiversity: TEEB for national and international policymakers
Umweltbundesamt (UBA) (2007) Ökonomische Bewertung von Umweltschäden. Methodenkonvention zur Schätzung externer Umweltkosten. Dessau
von der Pfordten D (1993) Sein, Sollen, Werten. Archiv für Rechts- und Sozialphilosophie 79:48 et seq
von Egan-Krieger T (2014) Die Illusion wertfreier Ökonomie. Eine Untersuchung der Normativität heterodoxer Theorien. Campus Verlag, Frankfurt a.M. et al
von Kutschera F (1982) Grundlagen der Ethik. De Gruyter, Berlin
Weimann J (1995) Umweltökonomik. Eine theorieorientierte Einführung. Springer, Berlin
Weimann J, Hoffmann S (2003) Warum Messung und Bewertung biologischer Vielfalt nicht unabhängig voneinander möglich sind. In: Weimann J, Hoffmann A, Hoffmann S (eds) Messung und ökonomische Bewertung von Biodiversität – Mission impossible? Metropolis, Marburg, p 17 et seq

Wieding J, Stubenrauch J, Ekardt F (2020) Human rights and precautionary principle: limits to geoengineering, SRM, and IPCC scenarios. Sustainability 12:8858

Williamson O (1993) Transaction cost economics meets Posnerian law and economics. J Inst Theor Econ 149:99 et seq

Wink R (2002) Generationengerechtigkeit im Zeitalter der Gentechnik. Evolutionär-institutionenökonomische Betrachtungen. Nomos-Verl.-Ges, Baden-Baden

Winter G (2001) Über Nutzen und Kosten der Effizienzregel im öffentlichen Recht. In: Gawel E (ed) Effizienz im Umweltrecht. Grundsatzfragen einer wirtschaftlichen Umweltnutzung aus rechts-, wirtschafts- und sozialwissenschaftlicher Sicht. Nomos-Verl.-Ges, Baden-Baden, p 97 et seq

Zangemeister C (2014) Nutzwertanalyse in der Systemtechnik. Eine Methodik zur multidimensionalen Bewertung und Auswahl von Projektalternativen, 5th edn. Winnemark

2 Idea-Historical Foundations and Dubious (Standard) Objections to Cost-Benefit Analysis

Abstract

Although economists do not experience their approach as normative, on closer examination it can be identified as an ethics based on economic science. Theoretically, it stands in the tradition of empiricism, which limits the realm of the knowable to facts (rather than of knowable norms as well), and if possible to reproducible and quantifiable facts. In this respect, lines in the history of philosophy can be drawn to cost-benefit analysis in twentieth century (and its economic foundation, so-called neoclassicism) already from medieval nominalism via the Reformation and Scottish moral philosophy. It would be too short-sighted to associate cost-benefit analysis solely with philosophical utilitarianism – with which there are some differences.

The objections raised against cost-benefit analysis in the tradition of the criticism of utilitarianism, such as the criticism of its orientation towards balancing, of an allegedly lack of distributive justice, or of the tendency to describe human beings as egoist in terms of behavioral science, are hardly convincing upon closer examination. In short, for various reasons, balancing is unavoidable in decision-making; absolutely balancing-free normative statements cannot be identified. Even the overestimated distinction between deontology and consequentialism (which is not feasible) does not change this. With respect to the behavioral description of human beings, on the other hand, economic research is more complex than sometimes assumed; and in general, this kind of description has also nothing directly to do with the (normative) cost-benefit analysis. Other objections against the cost-benefit analysis, which are not resounding on closer examination, can be named with the keywords quantification, informedness, concept of utility, individualism and criticism of capitalism or modernity.

In the debate as a whole, cost-benefit analysis and policy goal setting and balancing are linked in parts with the debate on the most effective instruments,

F. Ekardt, *Economic Evaluation, Cost-Benefit Analysis, Economic Ethics*, Environmental Humanities: Transformation, Governance, Ethics, Law 2, https://doi.org/10.1007/978-3-030-99284-2_2

e.g., of climate policy. For example, in the interest of climate protection, levies or environmental certificate markets can be conceived as instruments to internalize external costs. In the environmental context, economic policy instruments ultimately aim to induce people to do more to protect the environment by means of price signals. However, no one says that the desired limitation of environmental consumption must target a quantity that must first be determined by a cost-benefit analysis. In this respect, cost-benefit analysis and economic instruments are two separate objects.

2.1 Philosophical and Historical Foundations of an Empiricism of Normativity in the Form of Cost-Benefit Analysis

This chapter discusses some very common objections to the cost-benefit analysis that intend to fundamentally question this approach – but which, on closer inspection, may prove to be unconvincing (Sect. 2.2). This is preceded by a review of the historical basis of cost-benefit analysis in the field of classical empiricist thought in the history of philosophy (Sect. 2.1).

2.1.1 From Nominalism and Calvinism to Hobbes and the Beginnings of Epistemological Empiricism – First Highlight

It has already been noted that cost-benefit analysis ultimately leads back to questions about the origin of empiricist approaches. Empiricism as a movement in the history of thought is a very large topic, which can only be dealt with to a limited extent within the present limited framework. The focus here lies on determining the origins of a decision-making process oriented towards factual preferences of individuals, including the underlying idea that facts and precisely countable facts should be sought epistemologically. However, attention is not paid to the commonly made point already mentioned that concepts such as costs, benefits, and external costs have been introduced by authors such as Dupuit and Pigou since the mid-nineteenth century and the early twentieth century, respectively, and that there seem to be references to utilitarian ideas.[1] References to utilitarianism are being taken up briefly later, as soon as the nature of the cost-benefit-analytical balancing is understood and defended against partly inaccurate criticism (Sects. 2.2.3 and 5.2). From a philosophical-historical point of view, however, we will go back further here and ask about the more distant origins of the orientation towards balancing factual (and usually selfish) preferences as an ideal of rationality, including the secularism and individualism contained therein. The comparatively obvious view of the roots of

[1] With further references, e.g., Pearce et al. 2006, p. 32 et seq.

utilitarian approaches (such as Thomas Hobbes, David Hume, or Adam Smith) as topos already occasionally analyzed by others – although not very present in the discourse on cost-benefit analysis – is only briefly repeated here (Sect. 2.1.2). Primarily, the elsewhere almost ignored and even earlier roots of the history of ideas are to be taken up.[2]

In other disciplines, Thomas Hobbes is regarded by many as a modern thinker who is particularly associated with modern tendencies towards reason (in whatever sense), empiricism and individualism. The sharp boundaries between the modern philosophy of Hobbes, for example, and the pre-modern philosophies become blurred, however, when one draws attention to the aspect that even a legal, political, and moral theorist like Hobbes is influenced not only by socioeconomic factors but also by the history of ideas, even if he officially takes them up very little and, in the case of medieval aesthetics, even rejects them drastically.[3] In this book we will consider the influence of late scholastic nominalism, especially that of the Franciscan theologian and philosopher William of Occam, on the Hobbesian idea of a secular, individualistic and rational perspective on social issues – which breaks with Dominican-Thomist thought in various ways and redefines the relationship between knowledge and faith, reason and will, and man and God.[4] Nominalism directs, so it can be said at first in a rough outline, the gaze to the concrete, the individual, and to that which can be grasped with the senses. In this way it tries to separate between unprovable truths of faith and knowledge that can be understood with the senses. With this, and with the questioning of the subordination of philosophy and law to theology, a program is addressed that already comes a little bit close to the individualism-based avoidance of "irrational" decision-making programs, as it

[2] Marquard 1996, p. 370 et seq. suggests that, among other things, the attitude of Romanticism and anti-modernism, according to which the Middle Ages are the enlightened and radiant model from which modernity has fallen away because of the lack of medieval components, has contributed to the medieval/modern divide. Blumenberg 1966, p. 78 et seq. and passim advocates even more clearly for a resolution of this caesura, since the Middle Ages and the Modern Era are but two different responses to the same problem: the rejection of gnosis, the first response failing and thus necessitating the second. Such interpretations of the medieval-modern caesura stand in the context of the debate on medieval influences on modernity, which ultimately goes back to Max Weber; on Weber also Ekardt 2003, p. 99 et seq.; Ekardt 2001, §§ 14, 18.

[3] If one wants to speak of scholastic philosophy at all, because scholasticism actually refers only to the form of teaching (at the schola, the universities) and the method of theological (and philosophical) science at that time (the appeal to authorities, for example), but a great differentiation among the so-called scholastics can be discerned.

[4] Charles Sanders Peirce's view that the entire modern philosophy is based on Occamism seems to be one-sided, if only because many early modern philosophers had not studied Occam in detail; Leppin 2012, p. 273 et seq. Beckmann 2010, p. 172 et seq. Wieland, Entdeckung, passim, correctly points out that already in antiquity and especially in Aristotle there were some basic concepts which still live on in economics, but this will not be discussed in depth here. For a general discussion of the problems of this section, see Ekardt and Richter 2006, p. 552 et seq.

underlies the cost-benefit analysis as core intention, as can be pursued more closely below.[5]

The term nominalism describes in the narrower sense the theory of names of the general in the dispute over universals. This dispute is about whether the place of truth is to be found in things or in the statements about things – and whether general terms such as "the idea of a horse" enjoy some kind of reality or are merely a term for commonalities of real individual things. Here nominalists consider (a) universals as mere names (universalia sunt nomina) and know (b) only a truth of statements about things. Nominalism can also be understood as something more complex: as a logical, semantic, ontological and epistemological theoretical context. Logically, it is a theory which allows exclusively individual variables in the scientific discourse about things. Semantically this means, that general notions as signs denote something general (namely a more or less large number of single things), but do not denote anything general (because there is nothing in single things, which is general). Ontologically, nominalism as a theoretical complex is characterized by the notion that in reality there are only single entities. Finally, epistemologically, the nominalist theory is characterized by the fact that, because it is based on necessity and generality, knowledge and cognition are not gained solely by depicting structures of reality, but rather by dealing with purely (intra-)mental universal concepts in a way that is controlled by the rules of logic and grammar.[6]

As is well known, Plato was already concerned with the question of whether the individual or the general should be given more reality, and decided in favor of the general ideas in the world of ideas (the ideal horse), according to which individual things (the individual concrete horses) are derived and only imperfectly reproduced. For Aristotle, on the other hand, the general referred to the individual things existing in time and space in which it is incorporated. However, he – like Plato – stood by the opinion that one must grasp something of the essence of being (and not only of statements about things). Thus, he again referred to general contents that have more reality than the individual things. Neoplatonism and late antique Patristics follow this tradition of thought without further reflection on it.[7] It was not until early scholasticism that the question of the reality of general terms (universals) or individual things developed into the so-called universalia argument. As is well known, two basic views stood opposite each other: The first was the Platonic-Augustinian concept-realistic view that the general or general ideas have a higher reality than the individual things (universalia ante rem) – combined with the view that the place of truth lies in things. On the other hand, the view, first held by Johannes Roscellinus of Compiègne, that only the individual things are real and that the general ideas (universalia post rem) are nothing more than terms – which are then true as state-

---

[5] Thomas Aquinas did propagate a synthesis of theology and philosophy, but this ultimately insinuated a subordination of philosophy to theology. Ultimately, the principle that Thomas laid down in the Summa theologiae applied: Whatever in the other sciences proves to be in contradiction with theology must be rejected as false.

[6] See Beckmann 2010, p. 121 et seq.

[7] See on this in general Goldstein 1998.

ments about things only through their logical connection (so that the reference to reality is only in the individual things and not in the general ideas: the concrete individual horse has its correspondence in reality, not the general idea of the horse itself). The universals are thus mere names or, according to Roscellinus, word breaths (flatus vocis).[8] In this original discourse context, no explicit reflection on the recognizing subject central to later individualistic, rational and secular normativity has yet taken place. However, the latter already plays a role as the instance that determines the names out of its own interest and does not merely reflect the structure of things.[9] From the early fourteenth century onwards, nominalism took on a more concrete form faced with realism, as it was influenced by Thomas Aquinas. Johannes Duns Scotus, for example, taught, although he was a realist on the question of universality, that in each thing, in addition to the general nature (natura communis) or the general "what" (quiditas), there was a unique and special "this" (haecceitas), also in human beings.[10]

But it was only in the early fourteenth century that William of Occam articulated and pointed out nominalism in a way that was decisive for further development and also interesting for the context of later cost-benefit analysis. Occam initially opposed conceptual realism, since a general idea, unlike individual things, is not extramentally real.[11] For Occam only the individual is real, the universals are only possible predicates, which function as signs.[12] The general term (essence, being, etc.) is thus interpreted as a product of the cognitive mind and has no extramental reality of its own.[13] This view is related to Occam's economic principle, which has become so well-known under the metaphor Occam's razor (also Occam's razor): Never use

[8] Roscellinus explicitly applied his nominalist principles to the divine Trinity (as is clear in a letter to Abaelard), which thus no longer appeared as an identity; Goldstein 1998, p. 173 et seq. Roscellinus, however, recanted this heretical thesis at the Council of Rheims in 1092. Peter Abaelard, Roscellinus' disciple, attempted to reconcile the realist with the nominalist position shortly thereafter in his famous writing Sic et Non (1121/22). In the reality surrounding us, universals are only in things. For God they are before things, namely as archetypes of the created in his divine mind (in this respect he is a realist). And for humans they are indeed after the things, namely as concepts, which we must first deduct from the correspondence of the things (insofar he is nominalist).

[9] See Mensching 1992, p. 95.

[10] See Beckmann 2010, p. 108.

[11] "Quodlibet universale est intentio animae, quae secundum unam opinionem probabilem ab actu intelligendi non differt." (Every universal is an intention of the soul, which according to a probable opinion does not differ from the act of cognition). – Occam, Summa Logicae I, 15, 10.

[12] See Beckmann 2010, p. 126.

[13] "Quod enim nullum universale sit aliqua substantia extra animam exsistens, evidenter probari potest." (It can be shown with evidence that no universal is extramentally a substance.) Occam, Summa Logicae I, 15, 2. But this does not mean that universals have an (intra-)mental reality, for universals can be real (esse obiectivum) only currently in thought, for example, if one imagines the similarity relation between A and B just described in the mind. Occam assumes, however, that they are also permanently present in thought (esse subiectivum), even if they have been thought only once, and thus possess their own mental reality – as an available system of signs.

more (arguments, entities, etc.) than is necessary to substantiate a thesis.[14] The thus described tendency towards the individual and deflation of the realm of the recognizable results not only from Occam's nominalism (which focuses on individual things and statements about things) and "early criticism" but also from his voluntarism and here first of all from the doctrine of God's absolute freedom of will, which is directed "against the unity of a cosmic order and against the constraints of an unchangeable order of being given by God's will."[15] According to Occam there is no ratio creandi for God, i.e., no compelling reason for his creation, which is why he represents a kind of unfounded and all-founding will. The almighty, all-encompassing God thus becomes the powerful God of history, who created the world of his own free will and with his unlimited omnipotence (potentia dei absoluta) and could have created a quite different one. According to this omnipotence principle, God's action is not subject to any necessity, apart from the principle of freedom from contradiction.[16] On the other hand, God possesses the power to intervene in the order of salvation (potentia ordinata) whenever he wishes. The nominalistic God is thus a voluntaristic God. In contrast to Duns Scotus, Occam closely and radically links this freedom of God with the freedom of will of humans and contrasts the freely acting God with his likewise freely acting creature.[17] Through his freedom God sets human beings free – a freedom, which consists in his power to act. With this, Occam rejects the conventional idea of lex aeterna (the eternal law) and lex naturalis (the natural law) of Thomas Aquinas, i.e., the dominance of the cosmos over individuals, as well as the unchangeability of nature and cosmos. Occam sees clearly that it cannot be proven that the will really determines the actions of humans freely.[18] But free will is evident to him through experience. Freedom of will, early criticism and nominalism bring about at the same time a growing distance between the sphere of faith and that of knowledge – by strengthen-

[14] The principle of economy as a principle of thrift applies exclusively to human beings, since God is not bound to any economy by his comprehensive freedom, but human beings have to economize within the limits of their reason; Beckmann 2010, p. 43 and 46.

[15] Urban 1979, p. 34; also Blumenberg 1966, p. 169. Without very explicit reference to nominalism, it is further to be remembered in this context of Marsilius of Padua, who in some respects represents a bridge from Aristotelian to later Hobbesian views. In particular, Marsilius essentially knows only a procedural rightness of decisions, but not a substantive rightness; he is rather democratically oriented; and he knows rules only for the external behavior of people; on this, Miethke 2005, p. 38 et seq.

[16] Beckmann 2010, p. 36 und 40. According to this, everything (also God) is singularly existent and not conditioned by any necessity: „Nulla igitur substantia singularis est aliquod universale. Sed omnis substantia est una numero et singularis, quia omnis substantia vel est una res et non plures vel est plures res … Et ita vel erit processus in infinitum vel stabitur, quod nulla substantia est universalis ita, quod non singularis. Ex quo relinquitur, quod nulla substantia est universalis." – Occam, Summa Logicae, I, 15, 3.

[17] On this in more detail Urban 1979, p. 69. The notion recalls the immediate relationship (without intermediate instances and in pure experience/knowledge) of God and man in the reformation thought of, for example, Calvin, Luther, and Zwingli (who are indeed influenced by Occam); on this Ekardt 2003, p. 99 et seq.

[18] See, in summary, on the modern debate, Ekardt 2019, Ch. 3.2.

ing the individual and making God more transcendent. For humans this means that in the area of knowledge they must now rely on themselves (and their deflated reason); they become active as humans (and not as images of God that have already found the God-given world order and act according to divine principles in it).

In what has been said, as mentioned in Chap. 1, there is the idea of a scientific method that removes all unfounded metaphysical assumptions and considers only what can be made immediately understandable. The starting point is no longer the idea to be proven (or rather to be confirmed, since already given by the revelation), as in Thomas Aquinas, that in the end everything depends on God and is determined by him. With this skepticism towards general metaphysical evidence, however, human cognitive ability and thus the individual subject comes into play at the same time. Through the principle of economy and its reality, Ockham directs the view to the individual, the concrete, the subject only in the individual things.[19] But this tendency towards a critical examination and the nominalistic denial of a "physical" general present in the world also had the consequence that God – with whom Augustine still considered all ideas or general concepts – appears more and more transcendent. This, however, along with the emphasis on the individual, sets the course towards the secularization of human thought and action, which Occam hardly wanted. This is also confirmed by the fact that Occam takes the entire theology out of the realm of rational comprehensibility.[20] On the one hand, this ultimately means that God can only be believed in, but no longer known. And on the other hand, to the extent that the nominalistic God has begun to be the hidden one, all being loses its last necessity and thus gives way to the individual.[21] Thus, nominalism, early criticism, and the secularizing tendency resulting from both become causal factors for the stronger emphasis on the subject, as is characteristic of modern normative theories and the starting point of cost-benefit analysis. The tendency towards pluralism (again, probably partly unintentional) also points in this direction. This tendency is indirectly based on the orientation towards a critical examination in the forum of the individual – and it is also central to the later justification of cost-benefit analysis in the sense that a decision-making process must be established under pluralist conditions. Consequently, Occam writes, regardless of the only fragmentary character of his political-philosophical thoughts: The state represents a multiplicity of correspondingly constituted individuals, whose order is called politia, whereby the state authority, like e.g., the emperor, has its authority from the people.[22] State and community no longer appear here as a superordinate reality.[23]

[19] See, on the other hand, Mensching 1992, p. 95, on the comparatively still rather subjectless early scholasticism.

[20] Goldstein 1998, p. 296: "The increase of God becomes the condition of his overcoming." On the entire context also in parts Di Fabio and Schilling 2017.

[21] See Goldstein 1998, p. 122.

[22] See Occam, quoted in Beckmann 2010, p. 158 and 164.

[23] Beckmann 2010, p. 158. On the nominalist and Protestant features in the process of secularization, individualization and pluralization, see Di Fabio and Schilling 2017.

Some things about this development appear paradoxical inasmuch as the scholasticism of Thomas Aquinas in itself came with the claim to let reason increasingly enter human thinking. For Thomas, however, reason was rather the ability to see evidential certainties, which amounts to a substantialist and intuitionist, ultimately religious concept of reason. In this way of thinking, reason and thus natural law (lex naturalis) in legal and ethical questions do not stand for critically founded statements that prove to be tenable in critical discourses. Instead, traditionalisms disguised as evidence appear, in which some kind of subjective evidence experience masks the lack of reason (and thus the lack of rationality). Rational in the sense of scholastic natural law is then possibly the existence of God, the reality of the devil, the subordination of women to men, quartering of criminals, torture to find the truth in criminal proceedings, the fight against heresy, etc. Despite all the epochal relevance of a Thomas Aquinas, this only qualifies as a rational normativity in law and morality if one considers a substantialist reason to exist (i.e., a content-related list of propositions of reason, and not only an open reason as a mere decision-founding ability). It is, however, not evident that (until today) anyone would have been able to create such a list and make its fundamental existence plausible. By fighting the Thomist pseudo-reason, Ockham thus clears the way to a liberal criticalism that no longer assumes that more or less fatal necessities are essentially at work in world events but follows a discursive and procedural understanding of rationality. This is also shown by the fact that he cleans the question of universal truths or correctness from ontological aspects and makes it what it still is, for example, for a modern discourse theoretical rationalism: namely, the question of the truth and rightness of sentences.[24] Modern rationalism in particular would like to make it clear that such a rejection of a substantialist reason does not necessarily mean a rejection of a rational universalism in legal and moral questions in the sense of a skepticistic (ultimately declaring everything arbitrary or at least culturally relative) negation of globally right normative principles (discussed briefly below in Sect. 5.1). This insight into the – however limited – scope of reason in normative questions could be further developed into an exclusion of questions of good life from the area of what is rationally recognizable as well as into a doctrine of balancing in normative questions that depicts the boundaries of reason (briefly taken up below in Sects. 4.3 and 5.1). This insight into scope of reason could then later be further developed into an exclusion of questions of good life from the area of what is rationally recognizable as well as into a doctrine of balancing that depicts the boundaries of reason (Sects. 4.3 and 5.1). However, one could also try to reject the idea of a normative reason at all. This would be exactly the starting point of the cost-benefit analysis.

The world and, henceforth, also law and morality appear to be less static and compelling, so to speak. So, Occam assumes a voluntaristic God and thus apparently weakens the idea of reason by questioning the classical natural law embodying normative rationality in Thomas. .[25] As seen, the Thomist reason of classical natural law is ultimately no reason at all. Occam, with his doubt about the Thomist "reason-

[24] This might be interpreted as a precursor of a general transition from the philosophy of consciousness to the philosophy of language; on this Habermas 1981, Vol. 1, p. 518 et seq.; Böhler 1985, p. 29 et seq.

[25] On this view, expressed for instance by Blumenberg, see also Vossenkuhl 1990, p. 78 and passim.

able natural law," therefore does not necessarily weaken reason as such, but only a certain, untenable understanding of reason in legal and ethical questions. Occam does not clearly take the skepticistic position that rightness in normative questions cannot exist at all. But this is exactly the conclusion that Thomas Hobbes (see below) will later draw from Occam's preliminary work.

An essential bridge from the nominalism of Occam to Thomas Hobbes and the empirical tradition is formed by the approaches of the (above all Calvinist) reformers.[26] The reformation in the sixteenth century articulated a criticism of an authoritarianism and traditionalism of the official Catholic Church, both perceived as downright anti-Biblical. The revolutionary inspiration of the early Christian period, as the supporters of the reformation felt, had turned into a bureaucratically administered tradition that was enforced with official authority and coercion. A caste of "professionally religious" priests and monks was considered by the Catholic tradition to be closer to God through office. The Reformation negated all this; at its core, it placed people more strongly on themselves. The priesthood of all believers was propagated and the individual was referred directly to God. In addition to this relevant point for modern individualism, a focus on work – in the sense of making the divine calling visible – is characteristic of the emerging Protestantism, as is the striving to make one's own convictions real in the world. All this then also protected the modern ideal of growth, of technical progress, of the creation of some kind of paradise on earth. Religion stimulated economic and technological development and was at the same time under the influence of religion as well as the social emancipation of the early bourgeoisie. It drew its self-confidence from self-made economic success, in contrast to the supposedly lazy nobility. This Protestant influence has left deep marks on the actual emergence of liberal democracy and its liberal economic system.[27] This applies not least of all to the ambivalence of today's high standard of living and pronounced personal self-determination, and at the same time the threat of environmental destruction and poverty in other parts of the world. Individualism can hardly exist without the possibility of personal (also) economic development. Nevertheless, individualism, capitalism and also modern technology in the Christian discourse are by no means only affirmed, but have long been fought by certain circles.[28]

[26] On this bridge Ekardt 2019, Ch. 2.5; more detailed on the state of research on the Reformation MacCulloch 2004 as well as the contributions in Di Fabio and Schilling 2017.

[27] On the linkage also Welzel 2002, p. 34 et seq. and 237 et seq. (but without analyzing the ambivalences); Rosa et al. 2014, p. 57 et seq. and passim.

[28] The mutual influence of Calvinism and capitalism/state/human rights/democracy (Ekardt 2019, Ch. 2.5) is ignored by Scheidler 2015, p. 101 et seq. in Marxist manner. Likewise, it falls short if religious apocalyptic appears there solely as a consequence of economic and state violence, respectively as "opium for the people." It would be a common mistake, however, if one were to focus only on individual so-called intellectual greats in a description of the development of society as a whole (which is not the aim here!) and, moreover, were to fail to recognize the mostly not sudden, but rather organically flowing appearance of new ideas; on this, in detail, in the case of the Protestant Reformation, MacCulloch 2004, p. 27 et seq.

Like Occam and, after him, the Protestant-influenced Hobbes, the reformers, who then influenced Hobbes, were also seemingly skeptical of reason. First, they did not believe that humans could have a reasonable knowledge of God – for humans seemed to act only selfishly, because since Adam and Eve (due to original sin) they had been evil. They doubted reason in questions of value and metaphysics, and threw themselves entirely upon faith and predestination, that is, predestination for eternal salvation or damnation, in their relationship to a God who was thought to rule arbitrarily. At the same time, however, the closed medieval world view disintegrated with the individualization and focus on individual knowledge that lay, as well as pluralization. Moreover, the modern idea of democracy, which in turn was rooted in Calvinist concepts of community, and the Calvinist resistance struggle against religious oppression received their foundation in these lines of tradition. Individualism, personal freedom, and democracy, together with the striving for legal security and social upgrading of the emerging bourgeoisie, promoted an increasingly economically-oriented idea of freedom. Because the liberal tradition came soon to the insight that states are human-made and thus selfish (or "evil"). On this basis, a one-sided, anti-state and otherwise economically liberal concept of freedom soon developed. Consequently, freedom of property and freedom of religion were crystallization points in the struggle for liberal democracy and founded a one-sided view of freedom. At the same time, the (partly unintentional) roots of a liberal legal system become visible, in which the political institutions leave the individual in peace, both in a broad personal and economic sphere.[29]

Those reformation ideas and Occam's nominalism were taken up and continued by Thomas Hobbes in many respects, thus decisively shaping what can then be called the empiricist basis of cost-benefit analysis. For Hobbes, individual interests alone shape the order, which has ceased to be a transempirical, meta-historical quantity that can be derived from the principle of universality.[30] With Hobbes, the starting point is again the focus on the expectation of a (practically purely) selfish person. Without making a clear distinction between factual description of behavior and normative justification of behavior, Hobbes theoretically doubts more clearly than the reformers and more broadly than Occam that there can be any such thing as acting from a well-founded insight into the rightness of certain goals/values. Normative reason as a concept still seems compatible with Occam's ideas. Hobbes, by contrast, only knows interests or preferences and their balance among themselves, whereas the factual preferences are not judged as good or bad, as will be seen in the following.[31] Hobbes is well nominalistically of the opinion that there are no universals but only names, because without linguistic forms general ideas are not

[29] On this and the following, see Di Fabio and Schilling 2017; Ekardt 2003, p. 99 et seq. and 112; partly also MacCulloch 2004. For Augustine, the existence of state coercion and private property was justified by original sin, i.e., in modern terms, by self-interest; on the modern discussion of self-interest and altruism briefly below in Sect. 2.2.1 and in detail with further references in Ekardt 2019, Ch. 2.

[30] See Goldstein 1998, p. 267 et seq.

[31] See Hobbes 1651.

conceivable. For example, without language there would be neither the right nor the wrong; for right and wrong are exclusively linguistic attributes. He also defines good and bad completely differently than his predecessors did. He thus emphasizes Occam's theory by not only rejecting the thesis that there are universals outside of thought, but also by not recognizing any ideas that function as universals. According to Hobbes, the cause of thoughts lies in the perception of the outer and inner world of man through the senses. Hobbes says: I cannot think of any other activity of the human mind that would have been so naturally implanted in it that it requires nothing more than being born as a human being and being able to use his five senses. Sentiments are thus the natural basis of all imaginations that are built upon them, for "there is no conception in the human mind that was not first wholly or partially produced in the sense organs." All other conceptions are derived from sensations or sensation-based conceptions. The qualities that produce sensory stimuli are "nothing but different movements of matter." What distinguishes human beings from the animal is the fact of a special mental ability, which also refers to one's own conceptions and thoughts, by stringing together and linking names of things to affirmations, negations and other forms of language. A sequence of thoughts is then called mental discourse. Hobbes thus ultimately relies solely on experience, in accordance with the Galilean principles of a purely quantitative, mathematical and mechanical natural science, and attempts to apply its methods to law, morality and politics. This means that matter and motion are major components of Hobbes' theory of knowledge and, in a sense, constitute the material and effective cause of all other phenomena, whereby Hobbes categorizes all human action as striving or avoidance and thus justifies the empirical focus on exclusively conscious, calculated, usually selfish action, as it is the basis of cost-benefit analysis.

Hobbes cannot imagine anything that was or is not the object of sensation. It follows from this that one cannot have conceptions of universals because, for example, one cannot hear, see, feel, taste and smell the good, the one or the right and the wrong; they are linguistic attributes, that is, they require language to develop meaning. This is why Hobbes writes: Some scholastics say that the senses receive the species of things and transmit them to the center of the senses – a lot of words that explain nothing. Elsewhere he says that everything we imagine is finite and therefore there is no idea of what we call infinite.[32] Vice versa this means: Recording the consequences of our thoughts is the first purpose of language. Connected with this is its function to define names in a correct way: this is the beginning of all science. For with the onset of the ability to speak and the possibility of defining things, the ability to conclude can also come into play. For this reason, correct speaking for Hobbes initially requires only correct definitions. The truth of a statement then con-

[32] But how are the scholastic realists' ideas of universals to be explained according to Hobbes? Why can they conceive of universals? Hobbes would perhaps say that they are subject to an error because, starting from their sensations, they have wrongly inferred the causes of these and have then assigned names to them, just as they have assigned names to original sensations. Ideas, according to Hobbes, are "phantasms" that "are individual (in the same way) as that of which they are ideas"; see Esfeld 1995, p. 156.

sists in the correct arrangement of names, through which the meaning of the names acquires a coherent meaning. Because of this, the sovereign, in his or her legitimation by the people, is also given the authority to define terms. Hobbes' conception of language is thus directly related to his political theory since it leads directly to his idea that the sovereign can ultimately arrange arbitrary things and is not subject to any general principles (resulting from a substantialist or discourse-theoretical or otherwise understood normative reason). The sovereign, who has all-embracing competences like God, becomes the mortal God. The sovereign need not be a king, but can also be an elected majority. This understanding implies for law and ethics that either no more normative justification is possible or the way must lead to cost-benefit analysis with its building on factual individual preferences. Hobbes relies precisely on that, as we will see in more detail later.

Despite Hobbes' skepticism about the idea of normative principles of reason, it is not only in his anthropology but also in Hobbes' original nominalism that individualism becomes visible that makes him so influential for modernity. Hobbes divides names into proper names and common names (p. 26). Proper names are names that are semantically bound to a single thing. When a proper name is mentioned, only the idea of a specific thing or a specific person is evoked. But there are also names that could be given to many things that have certain things in common. Contrary to the realists, who would now claim that all men are essentially human beings and that humanity (in the sense of being essentially human) is included in the term man, Hobbes says that this name "man" is nevertheless the name of various individual things that no entity connects. For there is nothing general in the world except names, for everything named is individual and distinct (p. 26).

The nominalistic foundation of the cost-benefit analysis and its empiricist starting points with Hobbes can be illustrated by a more detailed analysis. It shows how Hobbes moves from his anthropology to the basic thesis of the theory of justice, namely that it is not possible to decide questions of norms, goals, and values based on reasons and that one must rather look at the factual preferences of people. This anthropology is in turn connected with its materialistic-nominalist epistemology. That manifests itself particularly clearly in Hobbes' linguistic analysis of a core concept of legal and ethical normativity: the concept of good. This concept, or the "good," was previously understood in many philosophies as an all-pervading entity and that which is to be striven for, and was considered universal in the Middle Ages. In contrast, Hobbes represents a subjectivist conception of the term good in contrast to the Thomist or Aristotelian tradition, which still assumed a substantial content of the good within the framework of its substantiationist conception of reason – and calls everything good that is desired and pursued. The Hobbesian view of happiness, which represents the constant success in attaining the things one desires from time to time (p. 48), is also connected with this. But humankind is by nature always in motion and therefore never completely satisfied. Happiness, in Hobbes' view, is also closely connected with the concept of power, in that power is the means of obtaining the goods by which one becomes blissful. Securing power contributes to the constant well-being which is the ultimate goal of bliss (p. 48). A lasting state of satisfaction is the goal, that is, the fewer things stand in the way in the attempt to obtain all the desired goods, the greater the happiness. It is thus the certainty that a state of satisfaction can be extended in content and time.[33] Consequently Hobbes denies the existence of a summum bonum (common good) that is somehow superior to individuals and also generally denies the existence of a finis ultimus or final goal. Rather, he particularizes the meaning of the term "well" and assigns to each individual the possibility to fill it with completely different content. The normatively correct is thus given a completely voluntaristic character and, as later in cost-benefit analysis, coincides with purely factual preferences. In contrast to Ockham, however, Hobbes clearly shows that he not only leaves the good life to each individual citizen, but also denies the possibility of universal, normatively rational principles of justice.

The translation of Hobbes' nominalistically informed anthropology into a theory of justice with the basic thesis "balancing the purely factual preferences of citizens" brings Hobbes himself into the picture of the state of nature.[34] For the conception of such a fictitious state of nature – the necessary avoidance of which is supposed to justify normativity – Hobbes starts from the individual human being. This individual person wants to and must fight for his or her survival without a state peace order. In doing so, Hobbes explicitly assumes that the individual human being may have any number of factual preferences; this then implies for him the fundamental freedom of the individual. Hobbes thus derives the idea of freedom, which for him is exhausted in the simple "absence of external obstacles," from purely empirical facts (i.e., from preferences). His approach is thus a typical example of the close methodological connection between metaphysical criticism and a narrow concept of freedom based on pure absence of compulsion. To be able to exist peacefully, humans must now put on artificial chains. Political freedom thus requires an act of subordination that is both obligation and freedom. Ultimately, Hobbes – like Occam before him – does not so much establish political absolutism as the rule of law or the state's monopoly on the use of force (whereby the lawmaker can also be an elected democratic majority) as a medium for resolving conflicts between the inevitably often colliding spheres of freedom of the citizens.

In the absence of the possibility of normative rationality, for Hobbes – is precisely that basic order just, which results when purely selfishly thinking (i.e., purely instrumentally rational) human beings have to come to an agreement (Hobbes assumes as a factual interest especially that of non-violence and stability). In this way, he anticipates the current positioning of cost-benefit analysts. This theory of justice is not only individualistic, because it only accepts the benefits of concrete people, and secular, because God is almost no longer present in this conception, and pluralistic, because it takes opposing views as an inescapable fact. Rather, it is also closely connected with Hobbesian anthropology as well as with his nominalism and voluntarism.[35] Hobbes' approach is thus directed against any approach that is perceived as metaphysical, and at the same time it is based as far as possible on the

[33] See Kersting 2002, p. 91.

[34] See on this also Macpherson 1962; Mahlmann 2010, p. 64 et seq. and 73 et seq.

[35] Instructive also the Hobbes summary in Hillgruber 1992, p. 5 et seq.

strictness of mathematics, which Hobbes adored.[36] Beyond the question of normative rationality, this concept of political freedom could be incomplete in essential points, since essential dimensions and points of reference of freedom such as the multipolarity of freedom, the preconditions for freedom, the democratic balance of powers, and a procedural concept etc. are simply missing. This, however, cannot be further elaborated in the present context (for a more diversified understanding of freedom, see Sect. 4.3 below). Nonetheless, Hobbes does state, covered by an absolutist terminology, that there should be an order that serves the factual individual preferences of all. And as such, he sees an order that has an absolute monopoly on the use of force and is limited (if possible) to preventing violence and property intervention, without, however, prescribing what a good life and the path to happiness is, which is also fully compatible with today's cost-benefit analysis.

2.1.2 From Hume to Smith and the Economic Classics – Second Highlight

Building on what has been shown so far, the immediate prehistory of cost-benefit analysis with Hume, the utilitarians and the economic classics will only be briefly described here. Because in this respect, it is a matter of facts already discussed to a greater extent by many voices in research. David Hume and later Adam Smith, who will be discussed in more detail in a moment, form a direct or indirect discourse context with their contemporaries Bernard de Mandeville, Francis Hutcheson and the Earl of Shaftesbury, which includes recourse to Thomas Hobbes and John Locke as intermediaries.[37] All authors mentioned have an anthropology, which considers purely factual ethical tendencies in humans, but without a separate normative level, against which to judge whether these purely factual feelings are then also right. It remains controversial between the authors whether humans are ultimately "purely selfish" on the anthropological level, i.e., whether supposedly altruistic feelings are ultimately only a concealed egoism (e.g., camouflaged vanity). Thus, Hume speaks of calm affects as the basis of ethics, which one could also cultivate and which would often help to formulate one's own concerns in an apparently objective language, but which nevertheless remain subjectively selfish feelings in the end.

In Hume's work,[38] Hobbes' concept of reason is ultimately found in an advanced form. Reason can identify suitable instruments and provide a factual basis, i.e.,

[36] Rühl 2005, p. 8 et seq.; on the later development Meyer 2003, p. 150 et seq.; Rogall 2013, p. 73 et seq. According to Hobbes, "universal" is simply something which, as a property, belongs to all things or persons to which it is referred. Universality, however, for Hobbes is exclusively a matter of extension, so it does not itself denote anything substantial: Universal is consequently a "name common to several things." For Occam, a universal as an intention of the soul or act of cognition is indeed something individual; but this is universal in the sense that it is a sign for several individuals. According to Hobbes, the name is something as individual as anything else being.

[37] Elaborated in more detail by Kopp 1995, p. 62 et seq.

[38] See for the following Hume, Treatise, Book III, Part I–III, each Section 1; aptly on Hume also Mahlmann 2010, p. 83 et seq.; Rohbeck 1978, p. 49 et seq. and 109 et seq.; on the German reception of the Scottish Enlightenment in detail Waszek 1988, p. 56 et seq.

subsumption material, for ethics and law, and compare comparable alternatives. Actual normative criteria, principles or preferences, which only express emotional drives of an egoistic or altruistically sympathetic nature, cannot be rationally caught up with. More than this is not possible, since an "ought" is not an object in the outside world and therefore cannot be verified. If morality were given a scientific (i.e., empirical for Hume) basis, this would lead to a conclusion from facts to norms. And quite independently of all this, a mere "ought" can never motivate anyone. All these views continue to aptly reflect the mainstream economic ideas and cost-benefit analysis that still prevail today, including the underlying empiricism.

The Scottish economist and moral philosopher Adam Smith is then considered the founder of the modern national economics.[39] His two main works on ethical feelings and wealth of nations made him famous far beyond the borders of Scotland.[40] In contrast to the mercantilists, he sees the source of the wealth of a country (with Locke) not in its precious metal stocks, but in the work done by the population, so to speak in good Calvinist tradition. The thesis that the division of labor promotes the productive forces functions, more than anything, as a basis of his welfare theory, emphasizing right from the start the economic importance he attaches to the division of labor. The larger the market, the greater the specialization of the individual and thus the extent of the division of labor. In determining the value of a good, Smith distinguishes between its utility value (the usefulness of the good) and its exchange value (the ability to use the good to acquire other goods in exchange). Both types of value can differ significantly in both directions, as the examples of water on the one hand and diamonds on the other show. Today, economists explain this with the theory of marginal utility, according to which, depending on the respective good, not every unit of every good has a value for the actor. Smith recognizes that market prices can fluctuate extremely depending on supply and demand. Furthermore, Smith emphatically advocates liberalization measures in action and vis-à-vis workers and farmers to make the human striving for self-interest as effective and unrestricted as possible, in his view for the benefit of all.

The focus on self-interest of (Hume and) Smith is also often overestimated. This is because the role of the ethical impartial observer and the sympathy of those authors is forgotten, while those are supposed to create the basic conditions for coordination among people without much regulation by the state.[41] Thus, in his theory of ethical feelings, in which he examines the origin and nature of moral judgment, Smith emphasizes that the pursuit of self-interest is tied to a moral of sympathy. Whether we judge a person's action as correct or wrong depends on whether we sympathize with the motive of the act. In his opinion, we imagine ourselves as the spectators of our own behavior and now we try to imagine what effect it would have

[39] Pars pro toto zu Smith Schumann 2005, p. 109 et seq.; Nutzinger 2006; Rühl 2005, p. 169 et seq.; Rogall 2015, p. 52 et seq.

[40] On the following Smith 1776.

[41] As an example, Kopp 1995, p. 386 et seq.; more clearly on Scottish empiricism Rohbeck 1978, p. 109 et seq. and passim; Rühl 2005, p. 169 et seq.; von Kutschera 1982, p. 148 et seq.; Rogall 2015, p. 52 et seq.

on us in this light. With Smith, this is the only mirror that allows us to examine the decency of our own behavior somewhat through the eyes of other people.[42] But Smith considers that people can be wrong and therefore expands the fiction to include the idea of an impartial observer who is not involved in the situation. The morality of sympathy, which is based on the ability of human beings to put themselves in other people's shoes, is thus, for Smith, a necessary prerequisite for self-interest to coincide with the common interest and for the individual to promote the welfare of all others by pursuing his individual interests. Having said all this, the approach remains committed to the empiricism of Hobbes or Hume by focusing on factual preferences in terms of altruism is postulated, so to speak.

Unlike modern economists, however, neither Hobbes, Hume nor Smith attempted to map these empiricist theoretical foundations through formalized methods, as one might have expected based on the natural-science-like ideal of knowledge since Hobbes.[43] This raises the question, which will be addressed later, whether the empirical observations by Hobbes et al. on human behavior are true (Sect. 2.2.1) – and even more so the question of whether their existence can normatively say anything (Sect. 5.1).

Nevertheless, by transferring natural scientific methods to economic processes and contexts, Smith gave economics an empiricist character that continues to shape them to this day. Such an orientation is initially also typical of natural science in the early modern period, although it was only with the advent of the neoclassical period that economics took on a comprehensive and mathematical character, including the now widely developed enthusiasm for models.[44] Moreover, 200 years ago, the French economist Jean-Baptiste Say articulated the idea that the value of a product is defined solely by its utility, which is expressed in the market price. In this sense, the emerging subjective theory of value states that the value of a good depends on its subjectively perceived utility. It could create demand through the production of goods, because it creates income, which creates the demand for products (Say's theorem). Thus, every supply creates a demand and both tend towards a state of equilibrium.

In the 1870s, the neoclassical period, founded by William Stanley Jevons, Léon Walras and Carl Menger, was developed accordingly in the field of economics, which cum grano salis has shaped economics to this day. The neoclassical economists now widely established the subjective value theory, which finally binds the value of goods completely to the preferences of the actors, without establishing any reference to the real price of, for example, the work used. Benefits – and also

[42] Smith 1759; Rogall 2015, p. 52 et seq.

[43] On this focus also in the epigones Schumann 2005, p. 118 et seq. Due to the unclear relation of morality and economics in Smith, there is, of course, also the thesis that modern empiricist economics with its formalism is precisely a rigid break with Smith; see Hühn 2015, p. 101 et seq.

[44] Lenhard and Carrier 2015, p. 10 et seq.; Luks 2013, p. 150 et seq.; Petersen and Faber, p. 172 et seq.; Hühn 2015, p. 101 et seq. A radicalization of ideas, but without mathematization, is found in Hayek 2014; in many points more differentiated Schumpeter 2005; exemplary for the empiricism also of other voices in literature Albert 2014, p. 75 et seq. and Baumgärtner 2012.

costs – are thus established as categories that are completely variable and can fluctuate greatly even for the same person depending on circumstances and temporal developments. In particular, the benefit of a good is negatively dependent on the quantity of the good available for the individual: The fewer units are available, the greater the benefit that a unit of the respective good provides (theory of diminishing marginal utility).

The subjective value theory of the neoclassical period is a central basis of the cost-benefit analysis, which is supposed to open up a maximization of the use in the sense of individual preference satisfaction. The other basis is a (normatively turned) conception of homo oeconomicus as a person who is always consciously and calculatingly oriented towards the benefit and its increase (and consequently cannot be influenced by others as well as one's own emotions and the situation), as it was taken over from the traditional line described above. It follows from this that the choice of the best possible alternative in each case amounts to maximizing the utility function under given secondary conditions, combined with the idea that this would lead to a complete balancing of all desires: The founder of the theory of general equilibrium was the French economist Léon Walras, who postulated a constant and complete balance of all preferences among each other and of supply and demand on markets, both on the goods market and on the labor and capital markets. As mentioned above, the first explicit beginnings of cost-benefit analysis are dated to the essay "De la Mesure de l'Utilité des Travaux Publics" by the French engineer Jules Dupuit, published in 1844.[45]

2.2 Why Common Criticisms of Cost-Benefit Analysis and Their Empiricist/Utilitarian Basis in Ethics Are Not Convincing

Based on the historical foundation of ideas, objections to the cost-benefit analysis are discussed below, which intend to fundamentally question this approach – which, however, on closer inspection prove to be unconvincing.

2.2.1 Homo Oeconomicus, Bounded Rationality, Behavioral Economics and the Blending of Descriptive Anthropology and Normative Theory

If someone is critical of neoclassical or related mainstream concepts of economics in any form, one usually quickly comes to homo oeconomicus, i.e., the concept of supposedly purely selfishly acting people. Thus, the idea seems plausible at first glance that cost-benefit analysis should also be seen as its expression and that the criticism of homo oeconomicus completely undermines cost-benefit analysis. In the history of ideas, it has already been noted that the thinkers who advocate an

[45] See Dupuit 1995, p. 55 et seq.

empirical approach to normative questions have anthropologically tended towards the figure known in economics as homo oeconomicus.

The concept of Homo oeconomicus in economics is an anthropological or behavioral description of human actions and omissions, which is, however, very much a pointed model, to explain economic constellations or future developments with a simple construct – the reaction to certain political control measures, for example. The concept in its current version implies a person who is always conscious, calculating, and oriented towards the benefit and its increase, with stable, consistent preferences and complete knowledge of all relevant information (and consequently cannot be influenced by others, his own emotions, and the situation).[46] One also speaks of rationally acting humans, which is adapted in other disciplines under the heading of rational choice approach. What exactly is meant by rationality is usually only vaguely determined; in particular, it remains vague whether rationality refers to goals and means in equal measure and whether the rationality of the ends shows itself, where possible, precisely in acting purely selfishly; this impression is, in any case, widespread. Homo oeconomicus appears as a model that does not necessarily have to reflect reality, but at least comes close to it. From an economic point of view, homo oeconomicus is woven into certain ideas of markets, especially into the general equilibrium theory already mentioned in the history of ideas, which assumes that if the actors are completely self-interested and informed, market activity leads to a situation in which supply and demand for economically tradable goods always balance each other, i.e., the market has a maximum problem-solving capacity for the aggregation of different preferences.[47] Homo oeconomicus often takes the form of statements that people generally behave in the form of markets or are guided by self-interest calculations of advantages and disadvantages, even in very private areas such as choosing a partner.[48]

The concept of Homo oeconomicus has met with widespread criticism, especially outside the field of economics, but also partly within the field of economics.[49] Homo oeconomicus is a frequent point of criticism in legal discourse, for example, because it is experienced as a contradiction to a liberal-democratic anthropology. As a possible object of criticism, there is not only the question of the empirical validity of the image of the constantly selfish, consciously acting and calculating human being (more about this immediately), but also the question of whether the concept of rationality used in this context is maybe limited from the outset (see Sect. 1.4). However, to make homo oeconomicus the main criticism of cost-benefit analysis is not convincing for two reasons:

---

[46] See critically Simon 1993; Saßmannshausen 2009, p. 61 et seq.; Schlaudt 2016, p. 24 et seq.; Aßländer 2015, p. 73 et seq.; Eidenmüller 2005, p. 28 et seq. and passim; Breunung 2001, p. 127 et seq.

[47] Schlaudt 2016, p. 18 et seq.; Eidenmüller 2005, p. 28 et seq.

[48] Classical pars pro toto Becker 1976.

[49] Aßländer 2015, p. 73 et seq.; Hühn 2015, p. 101 et seq.; Ötsch 2009, p. 107 et seq.; Saßmannshausen 2009, p. 61 et seq.; Rogall 2013, p. 110 et seq., 2012, p. 205 et seq.; Wurzer, p. 142 et seq.; Schlaudt 2016, p. 24 et seq.; in contrast to this: Schneider 2015, p. 55 et seq. and Marggraf 2001, p. 87 et seq.

The first reason is: For a long time now there have also been significantly more differentiated approaches within economics under headings such as bounded rationality or behavioral economics to explain human behavior in terms of behavioral science.[50] This is undoubtedly to be welcomed. Elsewhere, the phenomenon of climate change and the lack of societal reaction to it has been analyzed in terms of how companies, voters/consumers and politicians are often linked to each other in vicious circles – and how factors such as conceptions of normality, emotional perception problems with spatio-temporal distant consequences of one's own actions/convenience/habit are linked to them, self-interest, traditional values, technical-economic path dependencies and collective good structures have so far thwarted really meaningful climate protection efforts (and how aspects of all this that can perhaps be described as biologically conditioned interact with more cultural specifics in all these factors).[51] It has also been shown elsewhere that Marxist or Rousseauist-founded ideas that people are originally largely altruistic beings who only develop selfish drives through a capitalist environment are, despite their popularity in some circles, subject to clear doubts (and that the results of happiness research do not change these aspects).[52] The corresponding discussion will not be repeated here. The necessary differentiation in behavioral economics that has just been listed does not even reach the newest behavioral economics approaches completely; nevertheless, economic approaches to anthropology hit a relevant point with the reference to the human tendency to strive for self-interest, and economic approaches have been learning for a long time and are becoming more differentiated. Thus, economic theory formation on anthropology is often, but not always, self-interest-centered – rather, different perspectives stand side by side rather unconnectedly, even though some of them are really one-sided and often it is really only

---

[50] Compiled in Ekardt 2019, § 2 C.; fundamental, e.g., Simon 1955; Pearce et al. 2006, p. 283; Kirchgässner 2008, p. 202 et seq. and 232 et seq.; Simon 1993; Meyer 2003, p. 152 et seq.; Reckling 2003, p. 167 et seq.; Haller 2012, p. 39 et seq.; see also Klöhn 2006, passim; Rogall 2013, p. 110 et seq.; overseen in Neitzel and Welzer 2011, p. 16; Steinberg 2013, p. 101; furthermore on the more recent debate Ötsch 2009, p. 107 et seq.; Saßmannshausen 2009, p. 61 et seq.

[51] On this, see Ekardt 2019, Ch. 2, also on the question that in those factors parallel biologically and culturally grounded parts are visible. Thereby, deciding on an initial originality of "biology" or "culture" can be rather useless; see aptly Meggle 2009a, b, p. 285.

[52] Ekardt 2019, Ch. 2.6. Furthermore, it was important to prove there that despite the complexity of human motives and manifold obstacles, human learning ability is to be expected; classically in the twentieth century, e.g., Fromm 1996; Mannheim 1967, p. 237 et seq. and passim, Elias 1976, p. 43 et seq. and passim. All of this also includes a critique of behavioral science approaches that largely tie the explanation of individual and collective behavior to external factors, as exemplified by Kolb 2007, p. 72 et seq. and 274 et seq. It is certainly true when they state that the success of "social movements" can be related to the strength of the movement, its context, social fractures, changing social preferences, legal disputes, influences of international politics, and so on. But this seems trivial: of course, change (often) occurs because someone is striving for it, and people may be striving collectively, using existing institutions, to bring about change. However, this in no way explains the actual question of *why* social fractures, changed preferences, etc. occur. Thus, such statements are largely empty of content. Moreover, social movements – whatever this would mean – are of course only one element among many among the actors in a society.

about self-interest, sometimes even only about economic self-interest.[53] Already in view of the high complexity of world events, people are not completely sober mathematical finders of their goals and means in reality. All the more they are not always in the sense of selfish goals. And in spite of all real self-interest, the lack of stability and consistency of various preferences is also a constant obstacle to human beings, as even economists have found out.[54]

Nevertheless, against this background it is astonishing how strongly neoclassical economists believe that the purely model-theoretical, empirically incorrect homo-oeconomicus assumption has empirical forecasting power in behavioral research and thus also in governance research for policy instruments.[55] It is little reflected that this prognostic power is rather limited after the just said.

There is a second – more important – reason why homo oeconomicus is not suitable as a central point of criticism of cost-benefit analysis, as is already apparent from the preliminary remarks on epistemology (Sect. 1.4). It is wrong to imagine something normative under a behavioral model, i.e., a picture of how humans should be or what society should be like, which would mix anthropology and the theory of justice or ethics. In the worst case, the mixing results in a distorted perception of facts in the sense of a certain desired goal – or, conversely, in not being able to answer the question of how exactly a purely normative statement can be justified at all.[56] This error is equally widespread (though not universally) among economists and their critics.[57] Whatever human beings may behave in purely factual terms, for example, in climate protection, this nevertheless says nothing about how they ought to behave (ethically or even legally). Contrary to the typical view in legal discourse, homo oeconomicus cannot be compared with an alleged anthropology of liberal democratic constitutions because legal acts contain normative statements and do not make any empirical (including anthropological) statements on human behavioral drives. For this reason, the doctrine of the human image in German jurisprudence, contrary to common belief, does not make any statement about how human beings are, but rather a conception of what living together should be like. Namely, in such

[53] The latter, however, is not the case with Becker 1976, who is often criticized; in this respect, Neuhäuser 2011, p. 127 et seq. is correct (who mixes up the normative and the descriptive level, i.e. the distinction between economic evaluation and anthropology).

[54] Steinberg 2013; Weimann 2015, p. 114 et seq.; Elger 2011, p. 916 et seq.; Osterloh 2011, p. 932 et seq.; Rheinberg 2011, p. 929 et seq.; Simon 1993.

[55] As an example Wink 2002, p. 18.

[56] Exemplary, e.g., in Lenger 2009, p. 197 et seq.; Pinzler 2011, p. 99 et seq.; unclear Lenz 2015, p. 141 et seq.; Håller 2012, p. 39 et seq.; correct, however, Kirchgässner 2008, p. 45 et seq.; Endres and Holm-Müller 1998, p. 26 et seq.; van Aaken 2003, p. 89 et seq.; see also Krause 2010, p. 13 et seq.

[57] Therefore not correct: the rather critical voices (inside and outside economics) such as Helfrich 2011, p. 21 et seq.; Heinig 2008, p. 330 et seq.; Führ 2003, p. 257 et seq. and passim; Martini 2008, p. 169 et seq.; Honneth 2011, p. 323; Ostrom 1990, 2007; Spangenberg and Settele 2010, p. 327 et seq.; Binswanger 1998, p. 47 et seq.; Rothlin 1992; not apt, however, also the choice of words within economics itself by many authors; see exemplarily Hansjürgens and Lienhoop 2015, p. 65 et seq.; Sauer 2012; problematic in part also Di Fabio 2005; correct, however, Kirchgässner 2008, p. 45 et seq.; Endres and Holm-Müller 1998, p. 26 et seq.; van Aaken 2003, p. 89 et seq.

a way that there is a balancing of conflicting interests and no mere right of the selfish and the strong (see Sect. 4.3 for more details on the normative doctrine of balancing).

Regardless of all this, homo oeconomicus and cost-benefit analysis have strong points of common, which have already become visible above in the history of ideas, and which will repeatedly emerge in the following. The idea is in each case (once factual, once normative) that decisions are or (!) should be based on self-interest and completely calculated – and that human behavior is or (!) should be a balancing between alternatives under the boundary condition of the uncertainty of the other players' reaction. The reasons why it is not necessarily surprising for an empiricist approach to be able to reconcile both will be discussed repeatedly later on – namely when it becomes apparent that empiricist approaches such as cost-benefit analysis attempt to trace normative questions back to supposed factual questions.

It will not be pursued further here that in the economic approach to behavioral research, by focusing on reproducible experiments (which may be only limitedly suitable for the differentiated identification of human motives), a problematic empiricism can also be observed.[58] The focus in the present book is on decision theory, namely on cost-benefit analysis.

2.2.2 Criticism of Cost-Benefit Analysis as Normative and Therefore Unscientific and Subjective Theory

A further point of criticism of cost-benefit analysis (especially in the environmental discourse) aims at rejecting the idea of valuations as a scientific process in itself, because values are allegedly subjective per se and thus are not a possible object of scientific consideration. The actual assertion contained in this statement that valuations or statements on goals are subjective per se is discussed (and denied) in a separate section (Sect. 5.1). For the time being, the only point of interest here is that the point of criticism does not accurately capture the concern of the cost-benefit analysis. This is because the aim of cost-benefit analysis is to provide answers to normative questions in empirical form, so to speak, even though it will be shown (in Sect. 5.1) that this project cannot be carried out successfully. At the same time, it will be shown (in Sect. 5.1) that normativity is neither avoidable nor unscientific-subjective. Nevertheless, the objection is interesting here because it helps to clarify the normative character of cost-benefit analysis, as we will see immediately.

It should be noted that the purpose of cost-benefit analysis is not to make the trivial observation that deciding normative questions also requires knowledge of facts (which are obviously needed, e.g., as subsumption material to figure out whether the conditions of a legal or ethical norm are fulfilled in a given case). Cost-benefit analysis rather aims at making the normative standards as such empirical.

The basic idea of neoclassical (not only climate) economists that it is necessary to maximize the wealth of goods and to measure it in monetary terms, whereby the

[58] On this Ekardt 2021, § 1 D. III. 2.

purely factual preferences of individuals are unquestioned, is neither trivial nor can it be classified as a mere empirical concept. In contrast, the idea that precisely this view represents the optimal society is a normative concept. Correspondingly, it is being expressed from outside, but also within economics, that cost-benefit analysis is a kind of utilitarian ethics that replaces the factor of happiness with the factor willingness to pay and its balancing by means of the Kaldor Hicks criterion (Sect. 1.2) – *no matter how many economists may believe that they do not pursue any ethics.*[59] It is not really important here whether, given the difference between happiness and willingness to pay, one should not speak of an empirical rather than a utilitarian ethic (in the present book, this is indeed considered preferable). In any case, the thrust is identical, namely, to convert normative questions into factual questions and to answer them based on people's factual wishes. Beyond the references to the history of ideas, as they have already been dealt with, the normativity of cost-benefit analysis can be seen in several systematic aspects:

- The question raised in the cost-benefit analysis, which is the best possible degree of goal attainment of a multitude of possible goals and which goals are to be aimed at all, unspokenly addresses a valuation in the ethical sense, namely a statement about a desirable societal situation. Even if neoclassical economics avoids using the words fair, right, ought, or ethical, the talk of optimal or efficient conditions makes it clear that a cost-benefit analysis identifies options as preferable to other options. It is not by coincidence that one speaks of subjective value theory in economics, if the unquestionability of the individual preferences is postulated.
- In addition, it is obviously a normative requirement to believe that decision questions are to be clarified based on factual preferences of people, since this expresses a "should" in terms of how decision questions must be resolved. Economists call this consumer sovereignty.
- Furthermore, the (at least latent) equating of happiness in life with material prosperity is normative, which is recognizable in monetarization as well as in the talk of consumer sovereignty.
- Furthermore, if the factual preferences are shaped by the way they are determined in such a way that not only material prosperity but also a selfish perspective is distinguished, then this also has a normative character[60] (see in detail on the methodology of preference assessment Sects. 1.2 and 3.1). However, this does not necessarily happen in the cost-benefit analysis. It can also allow for altruistic preferences, even if it does not always do so.

Within the framework of the unacknowledged ethical approach of cost-benefit analysis, the aim is, as seen, what people prefer in purely factual terms – and what

[59] TEEB 2009; Lübbe 2009, p. 451 et seq., 2015, p. 24 et seq.; Eidenmüller 2005, p. 173 et seq.; Schlaudt 2016, p. 43 et seq.; Nestle 2012, p. 223 and passim; Fehling 2009, p. 97; van Aaken 2003, p. 90 et seq.; von Egan-Krieger 2014, p. 55 et seq.; Weimann 1995, p. 170 et seq.; Crane and Matten 2010, p. 98 et seq.

[60] Likewise also Myrdal 1976, p. 186; von Egan-Krieger and Ott 2006, p. 65; Wurzer 2014, p. 142 et seq.; von Egan-Krieger 2014.

results from this in monetary aggregation in society.[61] Quantifiable balancing thus appears as a symbol of rationality, which brings the preferences, which for their part cannot be rationally verified, into a single currency (money) and thus should make them comparable. Thus, when a neo-classical economist asks about the correct climate policy, he would not ask: Which normative climate policy framework for political decisions does political freedom and its preconditions together with certain balancing rules provide, including the rights of those living far away in space and time? Instead, economists would usually ask: How much would people living today pay for a stable global climate, or, expressed in market prices, what would be the advantages and disadvantages of climate change on the one hand and climate policy on the other? In other words, an economist would determine the subjective appreciation of individuals for certain consequences (assuming certain probabilities of those consequences) and derive from that how much climate protection the respective economists consider objectively correct or efficient.[62] Each further procedure, in particular a normative argumentation without figures is thereby identified as unscientific and irrational. The identification of science with facts and figures was already visible above in the analysis of the history of ideas on the predecessors of cost-benefit analysis in classical empiricism. In the present section, the aim was initially only to show that the undertaking cost-benefit analysis has a normative character; only later will it be discussed whether the undertaking appears to be tenable in terms of practical philosophy (Sect. 5.1), whether it can avoid unsolvable application problems (Chap. 3), and how it fits into the liberal democratic legal system (Chap. 4).

At the same time, what has been said further illustrates that normative cost-benefit analysis is based on a theory of action or anthropology (homo oeconomicus), as mentioned above, but can nevertheless be considered strictly separate from it. To put it bluntly, the difference can be summed up in the following simple formula: "People are purely factually benefit-oriented, perhaps even purely selfish" (anthropology), "and that is a good thing, and listening to people's purely factual – whether selfish or occasionally altruistic – preferences results in the optimal basic social order" (theory of justice). One could also say: Pareto optimality as well as Kaldor

[61] The manifold linguistic evasions of economists are traced in more detail by Lübbe 2009, p. 451 et seq. Some may object that cost-benefit analysis is not normative but explorative, which is then supposed to mean that it merely gives advice for the case of wanting to be rational. Such a third level of exploratory, however, does not stand between is and ought, but is again normative (and ergo not a separate level), for if its propositions are to be valid, it must also be willing to put its presuppositions-rationality-up for discussion. The entire objections of this thesis to cost-benefit analysis continue to apply unchanged either way – for cost-benefit analysis does not simply want rationality, but a very specific conception of rationality (namely, quantificational and preference-empirical). And it is precisely this that is shown in Chaps. 3, 4, and 5 to be illegal, inapplicable, and philosophically untenable. Nothing would change in the analysis of this volume, by the way, if instead of money another currency such as grains of rice or lifetime were used for quantification; apart from that, however, it is also difficult to imagine (even more than with money) how all costs and benefits could be translated into those currencies.

[62] Like this also, despite all their opposition, Stern 2009 and Nordhaus 2008, p. 38 et seq. and 59 et seq. That content of empiricist thought is precisely mentioned in Meggle 2010, p. 122 et seq.

Hicks criterion (Sect. 1.2) are based on a normative turn of homo oeconomicus.[63] So the point here is not to explain why anyone feels factually bound (or not bound) to normative propositions or how any normative propositions come about in fact. Due to their basis in utilitarian or empiricist ethics, cost-benefit analysts generally do not pursue these differentiations "descriptive explanation of behavior versus normative justification of behavior." Therefore, they assume that ethical aspects could be represented in cost-benefit analyses simply by measuring factual preferences for altruistic action.[64] The question (see Sect. 5.1) whether this is the right ethical perspective is not asked. The cost-benefit analysis depends, however, on the implicit affirmation of this question, because otherwise, in the absence of a robust theoretical basis, it could not identify optimal conditions, but could only describe sociologically factual conditions to a certain extent, which would not be the point of a decision procedure. For such a procedure should not describe factual decisions, but rather prescribe correct decisions.

Therefore, comparisons such as "efficiency versus justice" or "efficiency versus ethics and law," as both economists and legal critics maintain, are not convincing. They presuppose, contrary to what has been said, a non-normative character of cost-benefit analysis.[65] It is therefore a mistake when the IPCC examines ethics or the theory of justice as an add-on to the efficiency analysis. In the discourse, the contrast between ethics versus efficiency is in part more about the extent to which increased social distributive justice should take place for certain material goods in the sense of increased redistribution.[66] Whether this question can be posed in this way is not the focus of this book, but it will be dealt with at least briefly later on (Sect. 2.2.6).

The point that cost-benefit thinking is in reality itself normative and by no means analogous to the natural sciences was historically already emphatically addressed by Max Weber in the famous value judgment dispute of the economic and social sciences at the beginning of the twentieth century. However, the discussion at that time is characterized by the friction that the categories subjective and evaluative are more or less equated, so that any epistemological uncertainty related to economic issues makes this discipline seemingly evaluative (in contrast to what was shown in Sect. 1.4). Another misconception in this context was the idea that instrumental recommendations (for instance economic policy) are normative per se, although they initially reveal instrumental –descriptive – rationality in the form of if-then

---

[63] von Egan-Krieger 2014, p. 63 et seq.; Eidenmüller 2005, p. 28 et seq.; Pissarskoi 2014, p. 101 et seq. und 127 et seq.; Pearce et al. 2006, p. 32 et seq.

[64] See instead of many Wink 2002, p. 306 et seq.; Hansjürgens and Lienhoop 2015, p. 65 et seq.; Lange and Krull 2014, p. 137 et seq. (who, for example, simply paraphrase the criticism of cost-benefit analysis as a lack of factual acceptance and apparently see no difference from normative acceptability).

[65] In this respect, in my opinion, somewhat misleading therefore Nutzinger 2006; Grzeszick 2003, p. 647 et seq.; Mathis 2009.

[66] Exemplarily Panther 2006, p. 21 et seq.; Nutzinger 2006.

recommendations oriented toward specific goals.[67] Only these goals are normative, but not the if-then instruments that are proposed to achieve them. Contrary to some voices, human scientific knowledge is not normatively impregnated per se.[68] The argument in this regard that any descriptive explanatory approach is, in short, guided by an interest and that this interest, as a context of application, shapes the entire process of knowledge, is not convincing. For scientific explanation does not per se have to insinuate a practical use or, in short: a certain political position. Ultimately, the distinction between subjectivity and normativity is again not sufficiently made. Even the recommendation that certain governance instruments (e.g., levies) are more effective in achieving certain goals that are assumed to be normatively right (!) would not have a normative character, even if this is often misjudged.[69] In this respect, it is not the recommendation of certain policy instruments that is often associated with cost-benefit analyses that is of interest here, but merely the statement contained in the preference balancing that certain target states are more desirable or optimal than others.

The cost-benefit analysis is "empiricist" in the literal sense of "reducing knowledge to the experiential" (even if it turns out to be covertly normative) because it denies the possibility of objective normativity and thus ultimately takes a subjectivist or skeptical position on normative questions. Another variant of the reference to the empirical in the field of the normative is thus not chosen: There is no choice of any cultural historical tradition or "experience" as normatively welcome like in a culturally relativistic contextualism, communitarianism or Aristotelianism.[70] Likewise, there is no attempt to balance happiness as in utilitarianism. Rather, cost-benefit analysis recognizes only what can be experienced in the sense of what can be quantified and traced back to individuals, and thus the level of normativity is occupied solely by the current factual preferences of individuals. We will briefly return to these other approaches later (Sects. 5.1 and 5.2).

The problem of the cost-benefit analysis, which is thus more closely outlined in its normativity, does not lie in the fact that it would be reproached here for explicitly not wanting to be a theory of good life. This would be a theory which would standardize questions concerning only one person as well as the inner attitude of people (instead of merely conflicts of interests by external actions).[71] For such a theory, general standards are missing under the sign of liberal-democratic orders, so that such a theory cannot exist at all. In terms of good life, a liberal-democratic framework makes no ethical or legal stipulations; normatively, therefore, climate protec-

[67] Misjudged, e.g., in Suchanek 2000 as well as in various contributions in Jens and Romahn, Methodenpluralismus, passim; see Sect. 2.2.8 in detail on the distinction between normative cost-benefit analysis and descriptive debate on policy instruments.

[68] Inaccurate therefore Habermas 1968 and von Egan-Krieger 2014, p. 32 et seq.

[69] Too vague, e.g., Hansjürgens and Lienhoop 2015, p. 69.

[70] See aptly Brodocz 2007, p. 10; critical on contextualism Ekardt 2019, Ch. 3.1.

[71] At this point, caution is required insofar as some philosophers – e.g., Betzler and Schroth 2014, p. 279 et seq. – apparently understand this as something fundamentally different than the author; briefly commented in Sect. 2.2.3.

tion may only be formulated as the protection of other people, ergo as a conception of justice (Sect. 1.4), not as a kind of forced pacification of those who are restricted in their freedom for the sake of climate protection.[72]

Another misconception must be avoided: When in the following the hidden or misunderstood normative dimension in economics is addressed, this does not mean that a qualitative versus a quantitative dimension would be advocated. Qualitative research, for example, in the sense of qualitative-empirical expert interviews or individual case studies, may have its justification alongside quantitative empirical survey methods. But this qualitative perspective is not a normative perspective either, but establishes facts.[73]

2.2.3 Criticism of Balancing (or Consequentialism) as Such in the Tradition of Kantian Criticism of Utilitarianism

The connection of cost-benefit analysis with ethics in the empiricist or utilitarian tradition (see Sect. 5.2 for more details) makes it hardly surprising that another classic objection is raised against it: It is, at its core, the criticism of utilitarian approaches voiced in the tradition of Immanuel Kant and John Rawls. The essence of this criticism is that empiricist thinking inadmissibly allows for permanent balancing between different concerns.[74] This is seen as a violation of the principle of human dignity (which Rawls calls the principle of equal respect) or of the fact that certain things are intrinsically bad and therefore always forbidden. The observation that the cost-benefit analysis balances all interests against each other (by means of a monetarized balancing) is correct.

However, the idea that this is worthy of criticism or avoidable at all is not trivial. In any case, the demand for absolute normative statements should be kept free of a debate about the possibility of universal normative statements – valid everywhere and without cultural relativity (see Sect. 5.1), because universalism does not necessarily require "absolutism" but can accept balancing between different universal concerns.[75] Even those who, for example, regard the rights to freedom as universally justified may well allow balancing the freedom of one person with the freedom of another (and indeed must do so: see below and Sect. 4.3). We will also discuss sepa-

72 On this in general Hillgruber 1992; concretely based on sustainability and climate protection (including the fact that purely factually sustainability would of course nevertheless be more easily enforceable by other concepts of happiness, which however is precisely not a normative but a factual-enforcement-related consideration) Ekardt 2019, Ch. 3.4; on the descriptive side of the discussion, namely the question of whether people have (a) egoistic tendencies per se or only because of capitalist framework conditions and (b) what makes human happiness more or less likely, see with contrary viewpoints Hosang et al. 2005; Paech 2014; Ekardt 2019, Ch. 2.6.

73 On such methods Enders 2008, p. 75 et seq.

74 Rawls 1971; Betzler and Schroth 2014, p. 279 et seq.; Klie 2010, 103 et seq.; see also van Aaken 2003, p. 96 and passim; Martini 2008, p. 261 et seq.; Unmüßig 2014, p. 1.

75 For more details, see Ekardt 2019, Ch. 3.1–3.2, also on the further distinction universal versus global justice. The latter term is suitable to designate normative claims across borders (e.g., the question whether human rights not only apply universally – everywhere – but can also be asserted by citizens of one state against governments of other (!) states, if these are accused of having contributed to climate change, for example).

rately the distinction between deontology and consequentialism (on this at the end of the section) and the question whether persons may be quantified in balancing (Sect. 2.2.4).

In general, the criticism of balancing is not very suitable as the main point of criticism against cost-benefit analysis, because the persuasiveness of the objection is influenced by the basic ethical position one follows. If one follows the cost-benefit analysis, the balancing of the factual preferences and consequently their balancing against each other is very plausible (alternatively, one could think of a strict majority or consensus regime, which, however, would have to be justified first: see Sect. 5.1 for a short description). In this respect, the critique of trade-offs is in fact a questioning of the approach starting from factual preferences per se. This criticism will indeed be discussed later (Sect. 5.1).

Another argument against a critical focus on the paradigm of balancing is that one could ask whether, even if one tries to identify individual concerns as absolute, this does not at the same time implicitly contain the statement that opposing concerns have to take a back seat – i.e., that a balancing is carried out nevertheless. An even stronger argument against a critical focus on the issue of balancing is that not only in theories of preference, but also in a liberal-democratic legal system and a parallel ethical approach to the various human rights guarantees of freedom, balancing is unavoidable – and indeed genuine balancing with a result that is not fixed from the outset.[76] Any approach to ethics that denies this thus faces the challenge of having to justify the legal order of liberal democracies as unjust in central respects (on the problems this threatens, see Sect. 4.2).

Generally speaking, the system of liberal-democratic legal orders is a system that contains, at the basic constitutional level, rules of content and procedure for the collision of different spheres of freedom.[77] The concrete acts of parliament are then the representation of the balancing that becomes necessary in the process. In this context, liberal-democratic orders completely or largely avoid concerns that are not subject to balancing at all – all known constitutions stipulate the restrictability of freedom guarantees as the fundamental norms of liberal democracies (for ethical justification, see Sect. 5.1; for more details on balancing, see Sect. 4.3). If, for example, one wants to protect the climate for present and future people and, to do so, restricts today's automobile traffic, one protects freedom – but at the price of

---

[76] Accurately, e.g., as philosophers Meyer-Abich 2012, p. 376 et seq.; Francot 2014, p. 201 et seq.; as economists Gawel and Bretschneider 2012, p. 321 et seq.; Hansjürgens and Lienhoop 2015, p. 93 et seq.; Endres and Holm-Müller 1998, p. 24 et seq.; in legal discourse Fehling 2004, p. 443 (469); Ekardt 2019, Ch. 3.5–3.6; Susnjar 2010.

[77] Some of the questions that arise primarily in the legal (but also in the ethical) discourse have been discussed in more detail in Ekardt 2019, Ch. 3 with further references. For example, these are the questions of how freedoms, elementary preconditions for freedom and further conditions that promote freedom relate to each other; whether freedom from the state deserves more weight than the protection of freedom from one's fellow citizens; whether all concerns can be interpreted as an outflow of the principle of freedom; what understanding of human dignity as the basis of freedom lies behind this; whether freedom is also protected from only possible dangers; what the junction of freedom and responsibility for its consequences means, etc.

limitations of freedom in the present. The full realization of all spheres of freedom thus almost inevitably impedes each other to a greater or lesser extent. Likewise, the desire of Western companies for globalized, open markets without state regulations and taxes collides with the interest of many people in certain labor and social standards. Regarding this, general freedom of action (which includes mobility), freedom of property and the promotion of economic growth and jobs on the one hand, and the right to the elementary conditions of freedom life, health and subsistence on the other, are opposed. The inevitable trade-offs outlined in this way do not apply only to rare constellations or particularly dramatic problems such as climate change, but are literally permanently present in every public decision. By allowing the industrial society, approving industrial plants, permitting car traffic, etc., politics accepts statistically dead people, i.e., impairments of the right to the elementary prerequisites of freedom, due to the released air pollutants, etc. This happens in balancing with the freedom of consumption of all of us and with the economic freedom of the consumers. In this respect, one usually speaks camouflagingly of stochastic damages, i.e., statistical cases of illness and death, which occur at the latest in the long term and in combination with other causes of damage in the wake of the in-industrial form of life.

The preference approach can therefore not be criticized with the accusation, raised for instance by Rawls, that an empiricist ethics like utilitarianism – or indeed cost-benefit analysis – does not recognize absolute rights.[78] Indeed, the cost-benefit analysis does not, but this does not deserve criticism according to what has been said. The fact that Rawls arrives at a different result in this respect is essentially related to the structure of his two principles of justice, as has been discussed elsewhere.[79] These represent a small section of freedom that is only rarely (contrary to Rawls, of course, not quasi-never) affected by balancing in an absolute sense and formulate for all other conflicts of interest only the famous second Rawlsian principle of justice of the greatest possible advantage for the weakest. This is apparently supposed to mean that beyond this (already because of the ambiguities and empirical imponderables quite vague) requirement there is an unrestricted scope of action of the political authorities (for a instead differentiated balancing-theoretical approach, see below Sect. 4.3).[80]

Since balancing as such is unavoidable, it has been stated at length elsewhere that concrete decisions on sustainability (or on any other issue) cannot be guided by a principle of "do no harm" (neminem laedere); for most decisions this would be

---

[78] Rawls 1971; too little distinction between principles of justice and the balancing that follows is also made by Heinig 2008, p. 353 et seq. and Böckenförde 1991, p. 188 et seq.; for further criticisms of Rawls, see Ekardt 2021, § 3 C. and § 4 F. I.

[79] See in argument with Rawls 1971: Ekardt 2021, § 3 C. and 4 F. I.

[80] Kant, on the other hand, to whom Rawls implicitly refers, explicitly addresses the problem of reconcilable freedoms, but does not see any need for balancing. This is probably due to the assumption by Kant that it is clear where the freedom of one ends and that of the other begins. This has been discussed in more detail in Ekardt 2019, Ch. 3.5, 2021, § 3 C.

simply impossible.[81] If these balancing processes are decided by competent authorities, in particular by the democratic legislator when creating acts of parliament (i.e., abstract-general norms), it can happen that "exactly one" decision remains as the correct decision for the authorities and courts entrusted with applying the law to individual life situations. But in most cases, even for concrete individual decisions on the application of the law, there is still room for balancing.[82] Even if, for example, certain acts are prohibited by criminal law or certain environmental regulations exist for industrial plants under regulatory law, there is scope for legal interpretation with regard to legal concepts, scope for fact-finding and possible exceptions. And all of which are an expression of the conflicting normative concerns which are therefore precisely in a balancing relationship. The intrinsically "per se" prohibited acts, although sometimes postulated in philosophy, thus appear as essentially non-existent.[83]

It has already been analyzed in more detail elsewhere that even the reference to the principle of human dignity does not change the point that balancing is omnipresent (this is briefly repeated with reference to climate protection in Sect. 5.1).[84] In short, this resulted from the insight that the principle of dignity is not a norm applicable to individual cases, but rather the basic idea of liberal-democratic legal orders or the reason for the individual guarantees of freedom.

The hint that this would contradict our factual moral intuitions does not change anything (even if it were true).[85] For a conception of ethics based on factual moral intuitions proves to be unconvincing, as do related contextualist ethics or religiously based approaches. To avoid repetition of earlier work, we will only briefly discuss this later in the present book (Sect. 5.1).[86] In any case, this underscores the finding made at the beginning of the section that a balancing critique can ultimately only be read as a plea in the dispute about the quite generally correct approach to ethics.[87]

---

[81] Gawel 2001, p. 9 et seq.; Lübbe 2000, p. 73 et seq.; Winter 2001, p. 97 et seq.; Ekardt 2019, Ch. 3.5.

[82] This, as well as legal interpretation and the separation of powers in general, are considered briefly in Sect. 4.3 – and in more detail with further references in Ekardt 2019, Ch. 1.7 and 3.5.

[83] On this debate, for example, Betzler and Schroth 2014, p. 279 et seq.; partly also Lübbe 2016, p. 325 et seq.; classically Rawls 1971. Also, the example of Betzler and Schroth 2014, p. 279 (281), that one could not neglect one's own child in favor of two in Africa, is by no means an insight gained without balancing. For first of all (see above in the body text), there is indeed leeway for balancing in the application of that ethical and legal rule that is indeed to be recognized. And secondly, the creation of the rule is itself the expression of a balancing of various concerns, the result of which appeared to be the best outcome of establishing a duty of care for minors and entrusting them to their parents on a regular basis.

[84] Enders 1997; Ekardt 2019, Ch. 3.2; ignored by Honneth 2011, p. 151 and passim; Böckenförde 2003.

[85] Also on this Betzler and Schroth 2014, p. 279 et seq.; Lübbe 2016, p. 325 et seq.

[86] See in detail on various ethical approaches with further references Ekardt 2019, Ch. 3.1, 2021, § 3 B.-G.

[87] Conversely spoken, pointing out the constant necessity of consideration settles a number of arguments and approaches in many strands of philosophy. It shows, e.g., that there are no simple

At the same time, the reference to other conceptions of ethics brings up another point of view. Sometimes the criticism of the trade-off approach in cost-benefit analysis is also meant in a somewhat different way: namely as a statement in the time-honored philosophical dispute between deontological and consequentialist positions.[88] However, in spite of all tradition, such a distinction does not really seem to lead anywhere, because both positions cannot convince as clear positions. It has already been shown that the idea of consideration-free concerns, which is often associated with deontology, is not convincing. Conversely, there can be no genuine consequentialism without normative principles because consequences can always be judged based only on normative criteria.[89] In cost-benefit analysis, preferences are the criterion, and consequences are the degree of real fulfillment of those preferences. Major points of contention in the classical Deontology-Consequentialism controversy, moreover, tend to be aimed at an individual ethics, for example, the feared excessive responsibility and exact division of shares of responsibility in a consequentialist ethics. The underlying expectation of the possibility of a spelled-out individual ethics is, however, based, as has been shown elsewhere, on the assumption that there is not very much room for maneuver in concrete decisions and, accordingly, on ethical recommendations that can be clearly stated for individual cases and on the incorrect expectation of precise contributions to social problem solutions that can be concretized down to the individual.[90] Moreover, it is often ignored that any convincing ethical theory (no matter how it opts in the deontology-consequentialism debate) must justify statements about which agent is in charge of bringing about certain states of affairs.[91] For a clear limitation of duties as well as an order of responsibilities based on the rule of law are elementary for a liberal-democratic order (and that this should rather not be questioned is briefly considered in Sect. 5.1). One could also say: To overlook this has bad consequences, whereby the same could not be noticed without normative criteria like the rights to freedom

balancing-free commandments like the one not to lie per se: also behind this a balancing is hidden (the same applies to the prohibition of killing). Furthermore, a central analysis of so-called deontic terms such as commanded, forbidden, permitted then makes less sense than often assumed, because the difference "leeway versus no leeway" supposedly visible in these terms does not exist in such radicality (even if there are indeed, as also the theory of balancing – briefly on this in Sect. 4.3 – varying degrees of leeway to some extent). Also the focus on individual ethics becomes doubtful; briefly below in Sect. 2.2.7. Exemplarily all these points are visible in von Kutschera 1982.

[88] Most recently, Lübbe 2016, p. 325 et seq. (especially also on the question whether actions and consequences can be distinguished at all and where the demarcation might run); Betzler and Schroth 2014, p. 279 et seq.; Portmore 2011; Meyer 2014, p. 7 et seq.; classically von Kutschera 1982, p. 181 et seq., 63 et seq. and passim.

[89] Ekardt 2019, § 3 A.; in this direction also Meggle 2011, p. 23 et seq.; Portmore 2011.

[90] On this problems Ekardt 2019, Ch. 3.5–3.6.

[91] See on this Ekardt 2021, § 5 C. II. 3; Betzler and Schroth 2014, p. 279 et seq.; the point that not everyone is responsible for everything (and that liability and criminal law is therefore often less suitable for solving societal and namely environmental problems) is also a basic trait in Lübbe 1998.

(which are not free of balancing, see above). Thus, one cannot build a distinction between deontological and consequentialist positions on this basis either.[92]

2.2.4 Criticism of Quantification with Reference to Humans and Nature per se and of a Price for Environmental Goods

There is a further criticism that accuses cost-benefit analysis of partly basing the affectedness of people on the number of people affected and also of putting a price on nature by determining a monetary value for it. Some ethical theories find quantifying in terms of human life and health, in short the idea that fewer deaths are better than more deaths, objectionable per se (especially in medical ethics this is widely discussed).[93] The critique of quantification is obviously closely related to the critique of trade-offs as well as to the critique of consequentialism. Slightly modifying this, many postulate that nature, the climate, etc. should not receive a price.[94] This view (which is again reminiscent of the Kantian tradition) is justified either by the point that such a pricing is per se inadmissible, or by the argument that the destruction of natural goods is further accelerated in the case of a representation in quantified values.

However, this criticism seems unconvincing in several respects. The rejection in the following does not refer to justifying or rejecting the preference approach as such (see Chap. 5 below), and the rejection does not refer to discussing the problems with monetary quantification of goods that do not have a market price as such (see Chap. 4). Rather, the question to be discussed in this section is whether the idea of counting affected people, animals, etc. should be rejected per se – and whether the idea of assigning a monetary value to nature at all should be rejected per se.

First of all, balancing and quantification do not mean the same thing.[95] Balancing interests against each other does not necessarily imply an act of mathematization. Those who advocate such a rejection refer more or less clearly to the point that the intrinsic value of people and nature is allegedly disregarded if they are quantified

[92] Also in this regard, see the references just mentioned. An attempt to distinguish between the two can be found in Betzler and Schroth 2014, p. 279 (281 et seq.). It can lead to misunderstandings that there, in short, it is said that deontologists formulate a theory of the right and consequentialists a theory of the good, the former apparently being free of balancing and the latter open to balancing. Here the distinction justice versus good life is used terminologically for a completely different problem (by fixing the old Kantian distinction instead of the reference to freedom to the question of balancing). In the present case, justice simply means the area in which the interests of different people collide with each other, namely through external actions of persons. Questions of the good life, on the other hand, are those that concern someone alone, including questions about his inner attitudes.

[93] On this debate Taurek 1977, p. 293 et seq.; Parfit 1978, p. 285 et seq.; Rakowski 1991; Meyer 2006, p. 136 et seq.; Lübbe 2015; Otsuka 2004, p. 413 et seq.; Moellendorf 2014; Hanley and Barbier 2009, p. 322.

[94] Aldred 2006, p. 141 et seq.; Klie 2010, p. 103 et seq.; doubting Hanley and Barbier 2009, p. 320 et seq.

[95] See also Costanza 2014, p. 1.

and assigned a monetary value. The possible intrinsic value of nature will be discussed later in this section. As far as the intrinsic value of human beings is concerned, this is largely uncontroversial from an ethical and legal point of view. However, it is an open question whether the respect for the individual is undermined at all by quantifying life in particular. The economic evaluation of nature is also criticized as an expression of an exaggerated anthropocentrism and a mechanistic, linear understanding of nature. Humans know too little about nature to be able to understand its complex systemic interrelationships, which function in feedbacks, and thus to quantify them. In addition, nature must be recognized as a co-environment. Ecocentric positions, however, were rejected elsewhere as untenable.[96] Incidentally, it is unclear in an empirical sense whether the monetary representation of the environment in balancing on a psychical level increases or rather decreases the awareness of its importance. Economists hope for the former, critics fear the latter.[97] This does not appear to have been empirically clarified, although it will not be explored further here.

If, then it could be even more argued that the quantification of people in the sense of prioritizing the protection of the larger number of people is worthy of criticism.[98] Without going into the details of the intensive debate here, there are, however, clear doubts as to whether the assumed great importance of the question really exists (on the fact that, in any case, the reference to factual moral intuitions does not tend to lead anywhere, see Sects. 5.1 and 2.2.3). In general, the debate suffers from the fact that it essentially appears to be related to model situations between individuals that are not very close to reality, such as the fact that a drug that is definitely only available once is to be shared between two, three or four individuals. No matter whether cost-benefit analyses are applied to climate protection, to the health sector, to questions of biodiversity, or to still other constellations: In reality, there will always be a multitude of concerns and their impacts with varying degrees of uncertainty to be considered. If one were to completely ignore aspects of numbers, this would be tantamount to asking for completely arbitrary balancing. For if, for example, the mitigation of climate change for the sake of its economic consequences had to be treated on a par with various individual material desires of other persons (such as the desire to be allowed to continue to fly undisturbed in one's own helicopter for the purpose of preserving one's own enjoyment of life), simply because one does not want to impose any degradation, however understood, on the bearers of the singular special desires, the balancing would become largely arbitrary. This, however, cannot convince from any conceivable point of view, except from a decidedly skeptical one (against such a point of view see again Sect. 5.1).

In addition, it is not argued that the number is the only relevant aspect in balancing. It will be shown later (against the cost-benefit analysis) that balancing is an

---

[96] On this with further references Ekardt 2019, Ch. 3.4 – e.g., against Gorke 2000, p. 81 et seq.

[97] Aptly on this ambivalence Hanley and Barbier 2009, p. 326.

[98] On this Taurek 1977, p. 293 et seq.; Parfit 1978, p. 285 et seq.; Rakowski 1991; Meyer 2006, p. 136 et seq.; Lübbe 2015; Otsuka 2004, p. 413 et seq.; Moellendorf 2014; Hanley and Barbier 2009, p. 322.

ethical or legal decision within certain balancing rules.[99] (Sect. 4.3). One important rule is, for example, that the weight of the concerns in the specific case must be taken as a basis. However, this does not per se rule out consideration of the number of people affected. More than that is probably all that can be said about it. The real constellations are many times more complex than simple distribution constellations between three or four persons in some thought experiments, in which, moreover, (unlike climate change) reality is certain and anything but complex. Even if the cost-benefit analysis were considered convincing, it would not result in the feared simple counting of heads in the context of major social problems. When philosophers discuss whether a drug that could save one or five people from death should be given to one or to the five or whether, for example, a lottery should be drawn,[100] the right answer has to be: If one follows the balancing theory developed later, there is no clear commitment to one of those alternatives, because the basic principles of a liberal-democratic order and their implementation in balancing rules do not clearly specify any of the variants.[101] Against this background, it is not convincing to make quantification criticism as such a central point of criticism against cost-benefit analysis.

2.2.5 Criticism of Lack of Information – And to the Effect That Scientific Uncertainties Lead to Skewed Assessments on the Part of Scientists

If one enters into a conversation about cost-benefit analysis with natural scientists, for example at interdisciplinary conferences, and if these are natural scientists in contact with economists, such as biodiversity researchers, another point of criticism is formulated, as the author knows from multiple personal experience. The point of criticism usually assumes a common (!) evaluation process for, e.g., nature by natural scientists and economists and then regrets that already the natural scientists could only give uncertain evaluations because of the many uncertainties of their own natural scientific knowledge. Like the point of criticism in the previous section, this point breaks down into an aspect to be elaborated later and an incorrect aspect. The point that will indeed be portrayed as true later is that cost-benefit analyses can have great difficulties in concrete application because their collected factual material is subject to great uncertainties. For example, what good is information about the costs of more frequent hurricanes if I do not know for sure whether such

[99] In much more detail on balancing theory: Calliess 2002; Susnjar 2010, p. 199 et seq.; Ekardt 2019, Ch. 3.5–3.6.

[100] See on this Taurek 1977, p. 293 et seq.; Parfit 1978, p. 285 et seq.; Rakowski 1991; Meyer 2006, p. 136 et seq., 2014, p. 7 et seq; Lübbe 2015, passim; Otsuka 2004, p. 413 et seq.

[101] Ultimately left open also in Meyer 2014, p. 7 et seq. At the same time, however, the balancing rules also contain procedural rules that contain a responsibility for such decisions and ergo for the filling of the remaining leeway, ergo justify the power-separating and democratic institutional order; see very briefly Sect. 4.3 and in more detail Ekardt 2019, Ch. 3.5–3.7.

hurricanes will become more frequent because of climate change, nor do I know by how much they will become more frequent.

The inaccurate aspect of the criticism consists in the ideas that the facts are completely unmanageable and that the evaluation of desirable states of nature is a task for empirical researchers.[102] For example, there are comprehensive scientific studies on developments in climate change and biodiversity, which are also regularly bundled in reports such as those of the IPCC, so that both findings exist and can be easily read in bundled form. In this respect, it cannot be stated that the factual knowledge in central sectors that are particularly relevant to sustainability is not sufficient for action.

2.2.6 Criticism of the Concept of Utility as Well as Hidden Distributive Justice – True vs. False Utility and Intrinsic vs. Merely Instrumental Values

Criticism of the concept of utility as such is to be expected from many directions. Some might reproach the economic concept of utility for accepting arbitrary factual preferences of individuals without any content and for failing to differentiate between what might be called true utility and false utility. The latter would be, for example, in the intention of the objection, something like the consumption of the RTL television program. This objection must not be confused with the objections (to be discussed later) that the preference approach is linked to facts and seems to derive normative conclusions from them, and that there are possible problems in the detailed recording of preferences. The point of criticism here is rather that the notion of utility must be qualified in more detail from the outset, if necessary even against the expressed or implicitly discernible will of the participants.

Another further problem discussed in economics itself is that the best possible balance of costs and benefits seems to be indifferent to distributional issues.[103] This is true if the cost-benefit analysis is (commonly) understood in a way that only the sum of costs and benefits counts. Even if one makes clear (see already Sect. 1.4) that distributive justice is only one of the aspects of justice, one can therefore ask whether the pursuit of a sum of benefits is preferable to any other kind of distributive scheme, however understood.

As far as the criticism of the alleged lack of distinction between true and false utility is concerned, it must first be stated in legal terms that liberal-democratic legal systems do not make such a distinction. Rather, spheres of freedom of the individuals are defined, which then force balancing in case of conflict, without, however, treating certain wishes as welcome and others as per se disqualified from the outset. This legal approach is at least prima facie also convincing from an ethical point of

[102] See Pearce et al. 2006, p. 35; many economists also emphasize the complexity of modern societies, for example, Suchanek 2000, p. 71 et seq.

[103] Critical on this Panther 2006, p. 21 et seq.; Nutzinger 2006 and affirmatively Endres and Holm-Müller 1998, p. 22 et seq.; from a non-economic perspective Britz 1997, p. 185 et seq.

view, because otherwise a list of true and false benefits would be needed, for the compilation of which a concrete criterion would be lacking.[104] With its open concept of utility, cost-benefit analysis also avoids a problem that plays a role in philosophy in the debate on egalitarianism. There it is discussed, among other things, whether one can assume that, for example, a certain object has the same benefit for every human being – and what consequences it must have if this same benefit cannot be stated.[105] However, precisely this question (as well as the attempt to categorize true and false benefits) is likely to lead into insoluble problems. The point that liberal democracy does not prescribe any particular socio-political distributional order in terms of content and, moreover, has even completely excluded questions of the good life from the realm of what is accessible to legitimate regulation, is insofar the answer to another point. This is that, in view of a multitude of conceivable opposing arguments, there is no convincing criterion for the unambiguous grading of categories of benefit (including the above-mentioned disqualification of a supposedly false benefit). This debate, which has been conducted in depth elsewhere (and which should not be understood here as a rejection of balancing rules as outer limits: Sect. 4.3), will not be revisited here.[106] Whoever would like to attack the cost-benefit analysis of the benefit scheme or the indifference of distribution would have to assume that liberal democracy and its constitutional law have developed a philosophically unconvincing point of view in the last 200 years. According to the view represented here, this proof (and thus the proof of a criterion which makes utility qualifiable) ultimately does not succeed.

Consequently, the question of which distributive standards should apply has also been highly controversial in philosophy for a long time, which leads to very ramified detailed questions, positions and counter-positions in the debate on egalitarianism.[107] In the view expressed here (for the basic approach, see Sects. 4.3 and 5.1), there is no clear answer to such questions, not only constitutionally but also ethically. At the very least, however, they are not practicably applicable to the reality of the climate problem in particular, where life and death for a large number of people and, in general, many intertwined aspects are at stake. Accordingly, the complex social distributive justice debate will not be taken up in detail here.[108] However, even liberal-democratic legal systems do not give political actors the authority to put the physical foundations of liberal democracy as such at risk; in this respect, it is indeed possible to make essential decisions for climate protection that could not be made

[104] For the orientation towards "true" (and "false") freedom or utility, e.g., Böhler 1993, p. 201 et seq. and Murswiek 1994; in contrast, Ekardt 2019, Ch. 3.2.

[105] See in this regard in overview with further references Lerch 2006, p. 108 et seq.; Frankfurt 2015.

[106] On questions of justice and good life with further references Ekardt 2019, Ch. 3.4; Heinig 2008; Hillgruber 1992.

[107] On this, e.g., Frankfurt 2015; Rawls 1971; Sen 2009; Lerch 2006, p. 108 et seq.; Dietrich 2001, p. 132 et seq.; Dworkin 1981; Sennett 2003; on the empirical situation regarding distributional issues see Piketty 2014; Herlyn 2012, p. 17 et seq.

[108] On the legal debate with further references Heinig 2008 and Ekardt 2019, Ch. 3.4.

in this way by cost-benefit analysis (on this in Sects. 4.3 and 5.1), without having to accept egalitarian ideas, for example.

Another criticism on the topic of benefits could claim that there are different forms of benefits, some of which cost-benefit analysis tends to overlook. Economists are fixated on instrumental or non-intrinsic values, which means that their balancing decisions downplay essential (especially sustainability) concerns from the outset.[109] On the latter point, it should again be pointed out that the terminology used in the cost-benefit analysis uses utility and value synonymously and attaches both solely to people's subjective preferences – and that the starting point of economic thinking is generally economically relevant transactions. This does not mean, however, that people should necessarily express preferences only in relation to their income, for example, but only that utility and value are linked to the willingness to pay.[110] Indeed, it is to be questioned later (in Chaps. 4 and Sect. 5.1) that preferences are built upon at all and that their measurement in money is attempted. But it is not true that cost-benefit analysis from the outset focuses exclusively on economic products.

Another criticism could be that cost-benefit analysis is based on the idea that decisions always have disadvantages, i.e., costs. What is meant here is not the criticism of balancing in the sense discussed above, but a criticism to the effect that with intelligent use of technology it is inevitable from the outset that disadvantages of, for example, environmental policy decisions will occur at all. Such a position is advocated, for example, by Michael Braungart under the heading "Cradle to Cradle."[111] A perfect cycle management of natural resources is allegedly possible, and in general humans should no longer only avoid damages, but simply do good to each other exclusively. However, for the ecological questions Braungart wants to address, it seems empirically far-fetched to imagine that solutions exist that transform man into a living being that is exclusively useful for all areas of the environment. Not to be discussed separately here is the politically oblivious overestimation of technical solution options and in the last instance the almost religious implications of such an approach.[112]

2.2.7 Criticism of Individualism

Another objection, continuing the criticism of the concept of utility, revolves around the fact that economists could be reproached for the fact that economic evaluation always attributes values to individuals (and defends their consumer sovereignty and subjective evaluations). This is indeed the intention of the proponents of cost-benefit analysis, but they see this as a justification rather than a weakness of their approach.

---

[109] The misunderstanding can also be found in van Aaken 2003, p. 90 et seq.; Spangenberg 2017; on the economic discussion about utility see also Scholz 2011, p. 192 et seq.

[110] Endres and Holm-Müller 1998, p. 20; Hansjürgens and Lienhoop 2015, p. 90; as an example for the misunderstanding Martini 2008, p. 811.

[111] See Braungart 2014.

[112] See on this Ekardt 2014, 2021, §§ 1 B. III., 6 B.

They see this kind of individualism, in contrast to their allegedly purely empirical approach, as an initial value judgment, which, however, is generally accepted, especially in Western societies. In particular, no one should dictate to people what is good for them and what they should consume.[113]

It will indeed have to be examined later on whether the assumption of factual preferences as a normative core principle is convincingly justified (Sect. 5.1). Separate from this, however, is the criticism that an individualistic approach as such is to be criticized in normative theories. Both proponents and critics mix up several things in this regard that need to be kept apart:

- A critique (or affirmation) of individualism as such is likely to fail because the term is too vague to be seen as a precise concept to be welcomed or rejected.[114]
- If by individualism one automatically means an empiricist basic orientation of normative theories, then this criticism mixes up two points that should be separated. Non-empiricist approaches – and e.g., liberal-democratic constitutional law – can also be named individualistic, because they also take their starting point centrally from self-determination and individual rights of individuals. However, they do not necessarily justify this by referring to subjective preferences or an assumed consumer sovereignty. Thus, individualism does not equal empiricism.
- Moreover, individualism is not the same as unrestrained individualism in the sense of advocating a free play of forces without political regulation. It may well be that many economists support such ideas to a greater or lesser extent. However, such ideas are not necessarily linked to the cost-benefit analysis (for a differentiated balancing model see Sect. 4.3).

All these clarifications have nothing to do with allowing politics a kind of protection against oneself at the expense of the individual. On the contrary, the idea implied by cost-benefit analysis that this is inadmissible is not criticized here (see briefly Sect. 2.2.6).[115]

---

[113] Pars pro toto Endres and Holm-Müller 1998, p. V; Hayek 1933; on the tensions see also Hanley and Barbier 2009, p. 326.

[114] What could be meant is a starting point from the individual in the description of behavioral science (which is often negotiated under the unfortunate heading of methodological individualism). It could also mean a normative concept – which would be more obvious in view of the character of cost-benefit analysis. However, individualism could mean many things, e.g., a certain understanding of freedom, but which one exactly? For discussions of freedom as well as for the non-sensible juxtaposition of individual and structure in behavioral science, see with further references Ekardt 2019, Ch. 2 and 3.2; on the latter also Greve 2015; on the former also Morgenthaler 1999 and Calliess 2002.

[115] See Ekardt 2019, Ch. 3.4; Hillgruber 1992; more presupposed than justified and elaborated, e.g., in Rawls 1971 and Habermas 1992.

2.2.8 Criticism of Economic Policy Instruments – Necessarily Connected with (Criticism of) Economic Evaluation?

There is another objection that critics and (many) economists share, but some make this point as a criticism and some as a strength. The question is whether objections to economic policy instruments such as environmental levies or environmental certificate systems can be formulated as objections to cost-benefit analysis. This is precisely possible if economic policy instruments such as EU emissions trading or ecological taxes are necessarily based on an economic evaluation of the costs and benefits of climate protection. This is exactly what many assume, including economists and especially cost-benefit analysts.[116]

But this is by no means self-evident. In the debate as a whole, cost-benefit analysis and policy goal setting and balancing are linked relatively easily by many sides with the debate about the most effective instruments, e.g., of climate policy.[117] One could conceive of levies or environmental certificate markets as internalizing external costs. This means that the costs of environmental damage, which were previously imposed on the general public by the polluters, should be integrated into the market process and, if possible, charged to the polluter in accordance with the polluter-pays principle. This presupposes that the costs can be clearly quantified, especially in comparison to the benefits of environmentally harmful actions. If such a quantification is planned, this indeed implies a cost-benefit analysis. Here, as elsewhere, the example of climate policy is chosen, but one could also use examples from social or education policy, for example.

Going even further, one might even think that economic instruments not only require cost-benefit analyses as a method of goal setting, but that economic instruments are even part of those analyses. Indeed, an economic evaluation can be understood in such a way that the costs and benefits of political goals are examined, but at the same time their concrete implementation by certain instruments is examined. This is how (allegedly) the most "efficient" path of, e.g., climate policy, as desired by economists, emerges.

However, the twofold linkage shown here is by no means as compelling as many economists assume and as many critics then experience as a hook to attack economic evaluation and economic instrumentation as a whole under overall headings that are not further differentiated, such as that nature (or climate) should not have a price. The lack of differentiation here consists in the following: Economic policy instruments in the environmental context ultimately aim at quantity control of environmental resources. This is done either by artificially limiting the quantity of a

[116] See, for example, Lee 2004, p. 149 et seq.; Endres and Holm-Müller 1998 (e.g., first sentence of the blurb: economic evaluation as a necessary precondition of economic policy instruments); Martini 2008, p. 811; Weimann and Hoffmann 2003, p. 28 et seq.; Unmüßig 2014, p. 1; correctly, however, for example, Costanza 2014, p. 1; TAB 2015, p. 13; Ismer 2014, p. 67 et seq.

[117] In turn, there is confusion about the concept of economic policy instruments as such. Economic incentives for behavior are the common basic idea of economic instruments, but economic incentives also emanate from classic regulatory laws, since these also change the price situation for certain behaviors on the market. On those debates Ekardt et al. 2015; Ekardt 2019, Ch. 4.5.

resource or environmental damage (e.g., greenhouse gases) and then allowing the polluters to trade the remaining quantities (certificate approach), or by pricing environmental damage through levies or subsidies (or subsidy cuts) and thus indirectly achieving quantity control. However, no one says that the targeted quantity control has to target a quantity that has to be determined beforehand by a cost-benefit analysis. Just as well, the target can be found by what economists call a standard price approach or standard quantity approach.[118] In this context, a price for an environmental good or a consumable or non-consumable quantity is determined by a political decision (possibly underpinned by human rights: Sect. 4.3) and not by a cost-benefit analysis. For this very reason, the present considerations deliberately avoid common terms such as economization, monetarization or valorization of the environment and nature.[119]

It is also not true, as may be noted in passing, that economic instruments are necessarily linked to the anthropology of homo oeconomicus (irrespective of the fact that the latter, as shown in Sect. 2.2.1, is by no means necessarily linked to cost-benefit analysis). By no means do economic policy instruments respond only to human self-interest. They can, for example, also address conceptions of normality, as has been discussed in more detail elsewhere.[120] In any case, the sense of economic instruments is thus confirmed by their ecological effectiveness perspective; it does not need (or at least not primarily) the reference to their possible cost efficiency, and conversely, a rejection of the idea of cost efficiency would not deprive economic instruments of their basis. If the idea of avoiding governance problems (such as rebound effects, enforcement deficits, problems of depicting or shifting effects to other sectors or regions) is formulated positively, it can also be said that economic instruments alone – assuming a strict cap or a high price as well as a substantially and geographically broad application – have the potential to solve the core of the environmental problem as a problem of excessive resource and sink use. For these problems are quantity problems; and only economic instruments can reliably bring about absolute quantity reductions through caps or prices.[121]

---

[118] Rogall 2013, p. 413; Gawel 2001, p. 19 et seq.; Ekardt 2019, Ch. 4.5; Lange and Krull 2014, p. 180 et seq.

[119] See also Daly 2014; Costanza 2014, p. 1.

[120] See in detail with further references Ekardt 2019, Ch. 4.5–4.6.

[121] See in detail Weishaupt et al. 2020; Hennig 2017; Ekardt 2019, Ch. 4.5; von Bredow 2013, p. 353 et seq. Governance instruments are to be assessed (empirically) according to whether they effectively achieve a goal X or not; generalizing attributions such as "neoliberalism" tend to obstruct a sober and differentiated insight. Thus, there are doubts about the performance of regulatory (or even purely informational and self-regulatory) climate protection instruments due to rebound effects, shifting effects, enforcement problems, imageability problems, etc.; these doubts are of such a fundamental nature that they depend only very partially on the concrete design. In contrast, levies and certificate markets (possibly combined with border adjustments for imports and exports) can fundamentally address these governance problems.

2.2.9 Marxist Criticism of Capitalism and Conservative Criticism of Modernity – Argument Against Cost-Benefit Analysis?

Another objection could be that cost-benefit analysis can be understood as an expression of capitalist thinking, which conversely is understood by most economists precisely as a positive correlation.[122] In a broader sense, the criticism of capitalism also includes Jürgen Habermas's critique of what he calls the colonization of the Lebenswelt, according to which (among other things) economic ideas increasingly shape all areas of life.[123] In some respects, this argument follows Karl Polanyi, who criticizes the dominance of the market principle over society, work, nature and money as a whole and identifies it as the main evil of the twentieth century.[124]

However, it is unclear why cost-benefit analyses should only make sense under the rubric of capitalism. Ultimately, they could also be used as a basis for socialist five-year plans. This would even be very plausible, since cost-benefit analysis is originally an aid to state planning of all kinds. Otherwise, it is misunderstood that the market mechanism, with its orientation toward ability and willingness to pay, does not require a one-to-one transfer to political decision-making. Even if one (as one should sensibly do) distinguishes between the terms market economy and capitalism – i.e., in addition to the market principle, an order based on private property and committed to growth thinking – nothing changes in this assessment. This also leaves open the question of how great the influence of "capitalism" is on today's states of non-sustainability (a question to which a differentiated answer has been given elsewhere).[125]

Partly similar to the tendentially left-wing, capitalism-critical argument against cost-benefit approaches, by the way, a more conservative civilization critique of modernity could also be attempted.[126] However, this does not change the assessments we have just made.

Repetition Questions

1. What influence did nominalism have on the emergence of empiricism? (Sect. 2.1.1).
2. How was economic classicism formed from the ideas of Occam, Calvin and Hobbes? (Sects. 2.1.1 and 2.1.2).
3. Why is homo oeconomicus not the core problem of cost-benefit analysis, and what motivational factors are there all in all? (Sect. 2.2.1).
4. Why are normativity, balancing orientation and quantification not the central objections to cost-benefit analysis? (Sects. 2.2.2, 2.2.3, and 2.2.4).
5. Do the concept of utility, distributional issues and intrinsic values play a central role in the critique of cost-benefit analysis – and the critique of excessive individualism? (Sects. 2.2.6 and 2.2.7).
6. Why must the assessment of cost-benefit analysis and economic policy instruments be strictly distinguished from each other? (Sect. 2.2.8).
7. Can the criticism of cost-benefit analysis be formulated as a critique of capitalism and civilization? (Sect. 2.2.9).

References[127]

Albert H (2014) Nationalökonomie als Soziologie der kommerziellen Beziehungen, new edition. Mohr Siebeck, Tübingen

Aldred J (2006) Incommensurability and monetary valuation. Land Econ 82:141 et seq

Aßländer M (2015) Gier als Tugend? Der Homo oeconomicus und die Fiktion vom eigennützigen Handeln. Jahrbuch Ökonomie und Gesellschaft:73 et seq

Baumgärtner S (2012) Normative Begründung der Nachhaltigkeitsökonomik. In: Studierendeninitiative Greening the University (ed) Wissenschaft für nachhaltige Entwicklung! Multiperspektivische Beiträge zu einer verantwortungsbewussten Wissenschaft. Metropolis-Verlag, Tübingen, p 273 et seq

Becker G (1976) The economic approach to human behavior. University of Chicago Press, Chicago

Beckmann J (2010) Wilhelm von Ockham, 2nd edn. Beck, C H, Munich

Betzler M, Schroth J (2014) Konsequentialisierung – Königsweg oder Sackgasse für den Konsequentialismus? Z Philos Forsch 68:279 et seq

Binswanger H-C (1998) Die Glaubensgemeinschaft der Ökonomen. Marburg

Blumenberg H (1966) Die Legitimität der Neuzeit. Suhrkamp, Frankfurt a.M

Böckenförde E-W (1991) Staat, Verfassung, Demokratie. Suhrkamp, Frankfurt a.M

Böckenförde E-W (2003) Menschenwürde als normatives Prinzip. Juristenzeitung:809 et seq

Böhler D (1985) Rekonstruktive Pragmatik. Von der Bewusstseinsphilosophie zur Kommunikationsreflexion – Neubegründung der praktischen Wissenschaften und Philosophie. Suhrkamp, Frankfurt a.M

Böhler D (1993) Diskursethik und Menschenwürdegrundsatz zwischen Idealisierung und Erfolgsverantwortung. In: Apel K-O, Kettner M (eds) Zur Anwendung der Diskursethik in Politik, Recht und Wissenschaft, 2nd edn. Suhrkamp, Frankfurt a.M, p 201 et seq

Braungart M (2014) Cradle to Cradle. Einfach intelligent produzieren. Munich

Breunung L (2001) Umweltschutz unter dem Optimalitäts- und Effizienzpostulat. Die wohlfahrtsökonomische Konzeption in soziologischer Sicht. In: Gawel E (ed) Effizienz im Umweltrecht. Grundsatzfragen einer wirtschaftlichen Umweltnutzung aus rechts-, wirtschafts- und sozialwissenschaftlicher Sicht. Nomos, Baden-Baden, p 127 et seq

Britz G (1997) Umweltrecht im Spannungsverhältnis von ökonomischer Effizienz und Verfassungsrecht, Die Verwaltung:185 et seq.

Brodocz A (2007) Erfahrung – zur Rückkehr eines alten Arguments. In: Brodocz A (ed) Erfahrung als Argument. Zur Renaissance eines ideengeschichtlichen Grundbegriffs. Nomos, Baden-Baden, p 9 et seq

Brunnengräber A (2009) Die politische Ökonomie des Klimawandels. Oekom-Verl, Munich

Calliess C (2002) Rechtsstaat und Umweltstaat. Mohr Siebeck, Tübingen

Costanza R (2014) Commentary on monetizing nature. http://greattransition.org/commentary/robert-costanza-monetizing-nature-barbara-unmuessig

Crane A, Matten D (2010) Business ethics. Managing corporate citizenship and sustainability in the age of globalization, 3rd edn. Oxford University Press, Oxford

Daly H (2014) Commentary on monetizing nature. http://greattransition.org/commentary/herman-daly-monetizing-nature-barbara-unmuessig (02.01.2015)

Di Fabio U (2005) Die Kultur der Freiheit. Munich

Di Fabio U, Schilling J (eds) (2017) Weltwirkung der Reformation. Wie der Protestantismus unsere Welt verändert hat. Verlag C.H.Beck, Munich

Dietrich F (2001) Dimensionen der Verteilungsgerechtigkeit. De Gruyter, Stuttgart

[127] In accordance with legal practice, parliamentary, governmental and EU Commission documents as well as laws and judgments are not listed in the bibliography, as they can be found unequivocally based on the reference given in the continuous text or via the general search engines. The last access date for all Internet sources is 31.10.2021.

Dupuit J (1995) De la mesure de l'utilité des travaux publics (ursprünglich 1844). Revue française d'économie:55 et seq
Dworkin R (1981) What is equality? Part 2: equality of resources. Philos Public Aff II:194 et seq
Eidenmüller H (2005) Effizienz als Rechtsprinzip. Möglichkeiten und Grenzen der ökonomischen Analyse des Rechts, 3rd edn. Mohr Siebeck, Tübingen
Ekardt F (2001) Steuerungsdefizite im Umweltrecht. Ursachen unter besonderer Berücksichtigung des Naturschutzrechts und der Grundrechte. Zugleich zur Relevanz religiösen Säkularisats im öffentlichen Recht. Pro-Universitate-Verl, Sinzheim
Ekardt F (2003) Liberalismus, Besitzindividualismus und Handlungstheorie. Leipziger Universitätsverlag, Leipzig
Ekardt F (2014) Cradle to Cradle? Idealistisches Konzept mit totalitären Anklängen. Politische Ökologie 2:136 et seq
Ekardt F (2019) Sustainability. Transformation, governance, ethics, law. Springer, Dordrecht
Ekardt F (2021) Theorie der Nachhaltigkeit. Ethische, rechtliche, politische und transformative Zugänge – am Beispiel von Klimawandel, Ressourcenknappheit und Welthandel, 4th edn. (= 3rd ed. of the new edition). Baden-Baden
Ekardt F, Richter C (2006) Ockham, Hobbes und die Geburt der säkularen Normativität. Zur Genese von Säkularität, Individualität und Rationalität in Recht und Moral. Archiv der Rechts- und Sozialphilosophie 92:552 et seq
Ekardt F, Klinski S, Schomerus T (2015) Konzept für die Fortentwicklung des deutschen Klimaschutzrechts. Marburg
Elger C (2011) Der Spielraum des Individuums. Entscheidungsprozesse im Gehirn. Forschung Lehre:916 et seq
Elias N (1976) Über den Prozess der Zivilisation, Bd. II. Frankfurt a.M
Enders C (1997) Die Menschenwürde in der Verfassungsordnung. Zur Dogmatik des Art. 1 GG. Tübingen
Enders J (2008) Wissensnetzwerke in der Klimapolitik. Frankfurt a.M
Endres A, Holm-Müller K (1998) Die Bewertung von Umweltschäden. Theorie und Praxis sozioökonomischer Verfahren. Kohlhammer, Stuttgart et al
Esfeld M (1995) Mechanismus und Subjektivität in der Philosophie Thomas Hobbes. Frommann-holzboog, Stuttgart
Fehling M (2004) Kosten-Nutzen-Analysen als Maßstab für Verwaltungsentscheidungen. Verwaltungsarchiv:443 et seq
Fehling J (2009) Die Ethik des Value of a Statistical Life. Die Rolle individueller Risikokompetenz für die Legitimität des VSL. Rainer Hampp Verlag, Munich
Francot LMA (2014) Dealing with complexity, facing uncertainty. Morality and ethics in a complex society. Archiv für Rechts- und Sozialphilosophie 100:201 et seq
Frankfurt H (2015) On equality. Princeton University Press, Princeton
Fromm E (1996) To have or to be. London
Führ M (2003) Eigen-Verantwortung im Rechtsstaat. Berlin
Gawel E (2001) Ökonomische Effizienzforderungen und ihre juristische Rezeption. Ein problemstrukturierender Überblick. In: Gawel E (ed) Effizienz im Umweltrecht. Grundsatzfragen einer wirtschaftlichen Umweltnutzung aus rechts-, wirtschafts- und sozialwissenschaftlicher Sicht. Nomos, Baden-Baden, p 9 et seq
Gawel E, Bretschneider W (2012) Gehalt und Grenzen eines Rechts auf Wasser – ein Zwischenruf. Archiv des öffentlichen Rechts 137:321 et seq
Goldstein J (1998) Nominalismus und Moderne. Freiburg
Gorke M (2000) Die ethische Dimension des Artensterbens. In: Ott K, Gorke M (eds) Spektrum der Umweltethik. Metropolis-Verl, Marburg, p 81 et seq
Greve J (2015) Reduktiver Individualismus. Zum Programm und zur Rechtfertigung einer sozialtheoretischen Grundposition. Springer, Wiesbaden
Grzeszick B (2003) Lässt sich eine Verfassung kalkulieren? Juristenzeitung 58:647 et seq
Habermas J (1968) Erkenntnis und Interesse. Frankfurt a.M
Habermas J (1981) Theorie des kommunikativen Handelns, 2 Bde. Frankfurt a.M

Habermas J (1992) Faktizität und Geltung. Frankfurt a.M
Haller C (2012) Menschenbild und Wirtschaft. Eine philosophische Kritik und Erweiterung des Homo oeconomicus. Tectum Wissenschaftsverlag, Marburg
Hanley N, Barbier E (2009) Pricing nature. Cost-benefit analysis and environmental policy. Elgar, Cheltenham
Hansjürgens B, Lienhoop N (2015) Was uns die Natur wert ist. Potenziale ökonomischer Bewertung. Metropolis-Verlag, Marburg
Hayek FA (1933) The trend of economic thinking. London
Hayek FA (2014) The road to serfdom, new edition
Heinig HM (2008) Der Sozialstaat im Dienst der Freiheit. Zur Formel vom „sozialen" Staat in Art. 20 Abs. 1 GG. Tübingen
Helfrich S (2011) Commons/ Gemeingüter – ein Leitbild für das 21. Jahrhundert. Zeitschrift für Sozialökonomie 2:21 et seq
Hennig B (2017) Nachhaltige Landnutzung und Bioenergie – Ambivalenzen, Governance, Rechtsfragen. Metropolis, Marburg
Herlyn E (2012) Einkommensverteilungsbasierte Präferenz- und Koalitionsanalysen auf der Basis selbstähnlicher Equity-Lorenzkurven. Ein Beitrag zur Quantifizierung sozialer Nachhaltigkeit. Springer, Wiesbaden
Hillgruber C (1992) Der Schutz des Menschen vor sich selbst. F. Vahlen, Munich
Hobbes T (1651) Leviathan
Honneth A (2011) Das Recht der Freiheit. Frankfurt a.M
Hosang M, Fraenzle S, Markert B (2005) Die emotionale Matrix. Grundlagen für gesellschaftlichen Wandel und nachhaltige Innovation. Oekom-Verl, Munich
Hühn M (2015) Wie die Wirtschaftswissenschaften sich der Ethik entledigten. Jahrbuch Ökonomie und Gesellschaft:101 et seq
Ismer R (2014) Klimaschutz als Rechtsproblem. Steuerung durch Preisinstrumente vor dem Hintergrund einer parallelen Evolution von Klimaschutzregimes verschiedener Staaten. Mohr Siebeck, Tübingen
Kersting N (2002) Thomas Hobbes. Hamburg
Kirchgässner G (2008) Homo Oeconomicus. Das ökonomische Modell individuellen Verhaltens und seine Anwendung in den Wirtschafts- und Sozialwissenschaften, 3rd edn. Tübingen
Klie A (2010) Die Bewertung von Umweltgütern mittels Zahlungsbereitschaft. Woran Kosten-Nutzen-Analysen scheitern. GAIA:103 et seq
Klöhn L (2006) Kapitalmarkt, Spekulation und Behavioral Finance. Berlin
Kolb F (2007) Protest and opportunities. The political outcomes of social movements. Campus Verlag, Frankfurt a.M
Kopp T (1995) Die Entdeckung der Nationalökonomie in der schottischen Aufklärung. Natur- und sozialphilosophische Grundlagen der klassischen Wirtschaftslehre. Bamberg
Krause D (2010) Nutzen. Für Anwendungen entgrenzt und grenzenlos angewandt. Metropolis-Verl, Marburg
Lange C, Krull D (2014) Infrastrukturelle Wasserdienstleistungen. Zur Funktion kostendeckender Wasserpreise nach Art. 9 der EG-Wasserrahmenrichtlinie und ihre mögliche Anwendung auf den Aufstau von Gewässern als infrastrukturelle Wasserdienstleistung für Binnenschifffahrt und Wasserkraft. Metropolis-Verl, Marburg
Lee C (2004) Kritik der neoklassischen Umweltökonomik. Über die (Un-)Möglichkeit einer pareto-effizienten Umweltsteuer. Lang, Frankfurt a.M
Lenger A (2009) Gerechtigkeit und das Konzept des homo culturalis. In: Goldschmidt N, Nutzinger H (eds) Vom homo oeconomicus zum homo culturalis. Handlung und Verhalten in der Ökonomie. Lit, Münster, p 197 et seq
Lenhard J, Carrier M (2015) Mathematics as a tool. ZiF-Mitteilungen 2:10 et seq
Lenz C (2015) Zur Durchsetzungsfähigkeit von Suffizienzstrategien. Eine Analyse auf der Grundlage von Ansätzen der Neuen Politischen Ökonomie. Metropolis-Verlag, Marburg
Leppin V (2012) Wilhelm von Ockham. Gelehrter, Streiter, Bettelmönch, 2nd edn. Darmstadt

Lerch A (2006) Kann man Nachhaltigkeit ökonomisch begründen? Materielle Verteilungsfragen und Generationengerechtigkeit. In: Ekardt F (ed) Generationengerechtigkeit und Zukunftsfähigkeit. Philosophische, juristische, ökonomische, politologische und theologische Neuansätze in der Umwelt-, Sozial- und Wirtschaftspolitik. Münster, p 108 et seq
Lübbe W (1998) Verantwortung in komplexen kulturellen Prozessen. Breisgau Alber, Freiburg et al
Lübbe W (2000) Neminem laedere? Archiv für Rechts- und Sozialphilosophie. Beiheft 74:73 et seq
Lübbe W (2009) „Aus ökonomischer Sicht ..." Was ist der normative Anspruch gesundheitsökonomischer Evaluationen? In: Baurmann M, Lahno B (eds) Perspectives in moral science. Frankfurt a.M, p 451 et seq
Lübbe W (2015) Nonaggregationismus. Grundlagen einer Allokationsethik. Münster
Lübbe W (2016) Handlungen, Handlungskonsequenzen und das Vollständigkeitsaxiom. Ein handlungstheoretischer Kommentar zum entscheidungstheoretischen Konsequentialismus. Z Philos Forsch 70:325 et seq
Luks F (2013) Die Zukunft des Wachstums. Theoriegeschichte, Nachhaltigkeit und die Perspektiven einer neuen Wirtschaft, 2nd edn. Metropolis-Verl, Marburg
MacCulloch D (2004) Reformation – Europe's House Divided 1490-1700. The Folio Society, London
Macpherson CB (1962) The political theory of possessive individualism. From Hobbes to Locke. Toronto
Mahlmann M (2010) Rechtsphilosophie und Rechtstheorie. Nomos, Baden-Baden
Mannheim K (1967) Mensch und Gesellschaft im Zeitalter des Umbaus. Gehlen, Bad Homburg
Marggraf R (2001): Nutzungsunabhängiger Wert der Natur und ökonomisches Verhaltensmodell. Jahrbuch Ökologische Ökonomik:87 et seq
Marquard O (1996) Transformation und Kontinuität – Mittelalter und Moderne. In: Beckmann J (ed) Philosophie im Mittelalter. Hamburg, p 370 et seq
Martini M (2008) Der Markt als Instrument hoheitlicher Verteilungslenkung. Mohr Siebeck, Tübingen
Mathis K (2009) Efficiency instead of justice? Searching for the philosophical foundations of the economic analysis of law. Springer, Berlin
Meggle G (2009a) Sinn des Lebens. In: Jordan S, Nimtz C (eds) Lexikon Philosophie. Hundert Grundbegriffe. Reclam, Stuttgart, p 246 et seq
Meggle G (2009b) Ein paar Schrittchen zusammen. In: Grunwald S (ed) Festschrift für Sabine Rieckhoff zum 65. Geburtstag. R. Habelt, Bonn, p 285 et seq
Meggle G (2010) Handlungstheoretische Semantik. Berlin et al, De Gruyter
Meggle G (2011) Philosophische Interventionen. mentis Verlag, Paderborn
Mensching G (1992) Das Allgemeine und das Besondere. Der Ursprung des modernen Denkens im Mittelalter. J.B. Metzlersche Verlagsbuchhd, Stuttgart
Meyer M (2003) Überlegungen zur Rationalität institutionenökonomischer Modelle. In: van Aaken A, Schmidt-Lübbert S (eds) Beiträge zur ökonomischen Theorie im Öffentlichen Recht. Deutscher Universitätsverlag, Wiesbaden, p 149 et seq
Meyer K (2006) How to be consistent without saving the greater number. Philos Public Aff 34:136 et seq
Meyer K (2014) Was spricht gegen interpersonelle Aggregation? Z Philos Forsch 68:7 et seq
Meyer-Abich K-M (2012) Rechte der Natur zur kulturellen Wahrnehmung unserer natürlichen Mitwelt. Zeitschrift für Umweltpolitik und Umweltrecht 35:376 et seq
Miegel M (2010) Exit. Wohlstand ohne Wachstum. List Taschenbuch, Berlin
Miethke J (2005) Politiktheorie im Mittelalter – Thomas von Aquin, Marsilius von Padua, Lupold von Bebenburg. In: Bleek W, Lietzmann H (eds) Klassiker der Politikwissenschaft. C.H. Beck, cop, Munich, p 33 et seq
Moellendorf D (2014) The moral challenge of dangerous climate change. Values, poverty, and policy. Cambridge University Press, Cambridge, MA
Morgenthaler G (1999) Freiheit durch Gesetz. Mohr Siebeck, Tübingen

Murswiek D (1994) Freiheit und Umweltschutz aus juristischer Sicht, In: Kloepfer M (ed) Umweltstaat als Zukunft. Juristische, ökonomische und philosophische Aspekte – Ergebnisse des Ladenburger Kollegs „Umweltstaat". Economica-Verl, Bonn, p 55 et seq

Myrdal G (1976) Das politische Element in der nationalökonomischen Doktrinbildung. Bad Godesberg Verlag Neue Gesellschaft, Bonn

Neitzel S, Welzer H (2011) Soldaten. Protokolle vom Kämpfen, Töten und Sterben. Fischer, Frankfurt a.M

Nestle I (2012) The costs of climate change in the agricultural sector. A comparison of two calculation approaches, Dissertation, Flensburg

Neuhäuser C (2011) Unternehmen als moralische Akteure. Suhrkamp, Frankfurt a.M

Nordhaus W (2008) A question of balance. Weighing the options on global warming policies. Yale University Press, New Haven et al

Nutzinger H (ed) (2006) Gerechtigkeit in der Wirtschaft – Quadratur des Kreises? Metropolis-Verl, Marburg

Osterloh M (2011) „Homo Oeconomicus" versus „homunculus oeconomicus". Die Antriebe des Menschen aus Sicht der Psychologischen Ökonomik. Forschung Lehre:932 et seq

Ostrom E (1990) Governing the commons: the evolution of institutions for collective action. Cambridge University Press, Cambridge

Ostrom E (2007) Understanding knowledge as a commons: from theory to practice. MIT Press, Cambridge, MA

Ötsch WO (2009) Kognitive Grundlagen menschlichen Verhaltens. Kognitionswissenschaften und neoklassische Standardtheorie. In: Goldschmidt N, Nutzinger H (eds) Vom homo oeconomicus zum homo culturalis. Handlung und Verhalten in der Ökonomie. Lit, Münster, p 107 et seq

Otsuka M (2004) Skepticism about saving the greater number. Philos Public Aff 32:413 et seq

Paech N (2014) Liberation from excess. Munich

Panther S (2006) Gerechtigkeit in der Wirtschaft. In: Nutzinger H (ed) Gerechtigkeit in der Wirtschaft – Quadratur des Kreises? Metropolis-Verl, Marburg, p 21 et seq

Parfit D (1978) Innumerate ethics. Philos Public Aff 7:285 et seq

Pearce D, Atkinson G, Mourato S (2006) Cost-benefit analysis and the environment. Recent development. Paris

Piketty T (2014) Capital in the 21st century. Harvard University Press, Cambridge, MA

Pinzler P (2011) Immer mehr ist nicht genug! Vom Wachstumswahn zum Bruttosozialglück. Pantheon, Munich

Pissarskoi E (2014) Gesellschaftliche Wohlfahrt und Klimawandel. Umgang mit normativen Annahmen und Ungewissheiten bei der klimaökonomischen Politikberatung. Oekom-Verl, Munich

Polanyi K (1944) The great transformation

Portmore D (2011) Commonsense consequentialism. Wherein morality meets rationality. Oxford University Press, Oxford

Rakowski E (1991) Equal justice. Oxford

Rawls J (1971) A theory of justice. Cambridge University Press, Cambridge, MA

Reckling F (2003) Handlungskontexte und Handlungsrationalität – Annäherungsversuche zwischen Ökonomik und Hermeneutik. In: van Aaken A, Schmidt-Lübbert S (eds) Beiträge zur ökonomischen Theorie im Öffentlichen Recht. Deutscher Universitätsverlag, Wiesbaden, p 167 et seq

Rheinberg F (2011) Vom Motiv zur Motivation. Über äußere Anreize, passende Belohnungen und optimale Anregungsbedingungen. Forschung & Lehre:929 et seq

Rogall H (2012) Nachhaltige Ökonomie. Ökonomische Theorie und Praxis einer Nachhaltigen Entwicklung, 2nd edn. Marburg (3rd ed. angekündigt für 2018)

Rogall H (2013) Volkswirtschaftslehre für Sozialwissenschaftler. Einführung in eine zukunftsfähige Wirtschaftslehre, 2nd edn. Springer, Wiesbaden

Rogall H (2015) Grundlagen einer nachhaltigen Wirtschaftslehre. Volkswirtschaftslehre für die Studierenden des 21. Jahrhunderts, 2nd edn. Marburg

Rohbeck J (1978) Egoismus und Sympathie. David Humes Gesellschafts- und Erkenntnistheorie. Campus-Verlag, Frankfurt a.M. et al
Rosa H et al (2014) Weil Kapitalismus sich ändern muss. Springer, Wiesbaden
Rothlin S (1992) Gerechtigkeit in Freiheit – Darstellung und kritische Würdigung des Begriffs der Gerechtigkeit im Denken von Friedrich August von Hayek. Frankfurt a.M
Rühl UFH (2005) Moralischer Sinn und Sympathie. Der Denkweg der schottischen Aufklärung in der Moral- und Rechtsphilosophie. Mentis, Paderborn
Sachs W (2013) Missdeuteter Vordenker. Karl Polanyi und seine „Great Transformation". Politische Ökologie 133:18 et seq
Saßmannshausen SP (2009) Der homo oeconomicus im Spiegel kognitions- und biopsychologischer Erkenntnisse. In: Goldschmidt N, Nutzinger H (eds) Vom homo oeconomicus zum hmo culturalis. Handlung und Verhalten in der Ökonomie. Münster, p 61 et seq
Sauer T (2012) Elemente einer kontextuellen Ökonomie der Nachhaltigkeit. Der Beitrag Elinor Ostroms. In: Sauer T (ed) Ökonomie der Nachhaltigkeit. Grundlagen, Indikatoren, Strategien. Metropolis-Verl, Marburg, p 135 et seq
Scheidler F (2015) Das Ende der Megamaschine. Geschichte einer scheiternden Zivilisation. Promedia, Wien
Schlaudt O (2016) Wirtschaft im Kontext. Eine Einführung in die Philosophie der Wirtschaftswissenschaften in Zeiten des Umbruchs. Vittorio Klostermann GmbH, Frankfurt a.M
Schneider D (2015) Wissenschaftsgeschichtliches gegen ein „Akteureskonzept des Homo oeconomicus". Jahrbuch Ökonomie und Gesellschaft:55 et seq
Scholz R (2011) Environmental literacy in science and society. From knowledge to decisions. Cambridge University Press, Cambridge
Schumann O (2005) Wirtschaftsethik und Politische Ökonomie in theoriegeschichtlicher Perspektive. In: Beschorner T, Eger T (eds) Das Ethische in der Ökonomie. Festschrift für Hans Nutzinger. Metropolis-Verl, Marburg, p 109 et seq
Schumpeter JA (2005) Kapitalismus, Sozialismus und Demokratie, 8th edn. Tübingen et al
Sen A (2009) The idea of justice. Harvard University Press
Sennett R (2003) Respect in a world of inequality. W.W. Norton, New York
Simon H (1955) A behavioral model of rational choice. Q J Econ:99 et seq
Simon H (1993) Homo rationalis. Frankfurt a.M
Smith A (1759) The theory of moral sentiments
Smith A (1776) The wealth of nations
Spangenberg J (2016–2017) Mehr wert als der Preis – Werte, externe Kosten und warum ihre Berechnung in Geldeinheiten mehr schadet als nutzt. Jahrbuch Nachhaltige Ökonomie:309 et seq
Spangenberg J, Settele J (2010) Precisely incorrect? Monetising the value of ecosystem services. Ecol Complex:327 et seq
Steinberg R (2013) Die Repräsentation des Volkes. Menschenbild und demokratisches Regierungssystem. Nomos, Baden-Baden
Stern N (2009) A blueprint for a safer planet. How to manage climate change and create a new era of progress and prosperity. Bodley Head, Cambridge
Suchanek A (2000) Normative Umweltökonomik. Tübingen, Mohr Siebeck
Susnjar D (2010) Proportionality, fundamental rights, and balance of powers. Martinus Nijhoff Publishers, Leiden
TAB (Büro für Technikfolgenabschätzung beim Deutschen Bundestag) (2015) Inwertsetzung von Biodiversität. Wissenschaftliche Grundlagen und politische Perspektiven. Berlin
Taurek J (1977) Should the numbers count? Philos Public Aff:293 et seq
TEEB (2009) The economics of ecosystems and biodiversity: TEEB for national and international policymakers
Unmüßig B (2014) Monetizing nature – taking precaution on a slippery slope. http://us.boell.org/2014/08/26/monetizing-nature-taking-precaution-slippery-slope
Urban C (1979) Nominalismus im Naturrecht: Zur historischen Dialektik des Freiheitsverstandnisses in der Theologie. Stuttgart

van Aaken A (2003) Normative Grundlagen der ökonomischen Theorie im öffentlichen Recht. In: van Aaken A, Schmidt-Lübbert S (eds) Beiträge zur ökonomischen Theorie im Öffentlichen Recht. Deutscher Universitätsverlag, Wiesbaden, p 89 et seq

von Bredow H (2013) Energieeffizienz als Rechts- und Steuerungsproblem. Unter besonderer Berücksichtigung der erneuerbaren Energien. Metropolis-Verl, Marburg

von Egan-Krieger T (2014) Die Illusion wertfreier Ökonomie. Eine Untersuchung der Normativität heterodoxer Theorien. Campus Verlag, Frankfurt a.M. et al

von Egan-Krieger T, Ott K (2006) Kritik der Neoklassik als Grundproblem der Wirtschaftsethik. In: Dürmeier T, von Egan-Krieger T, Peukert H (eds) Die Scheuklappen der Wirtschaftswissenschaft. Postautistische Ökonomik für eine pluralistische Wirtschaftslehre. Metropolis-Verl, Marburg, p 63 et seq

von Kutschera F (1982) Grundlagen der Ethik. De Gruyter, Berlin

Vossenkuhl W (1990) Ockhams Verständnis der Schöpfung. In: Vossenkuhl W, Schönberger R (eds) Die Gegenwart Ockhams. Munich, p 78 et seq

Waszek N (1988) The Scottish enlightenment and Hegel's account of "Civil Society". Springer, Dordrecht

Weimann J (1995) Umweltökonomik. Eine theorieorientierte Einführung. Springer, Berlin

Weimann J (2015) Verhaltensökonomik und Politikberatung, Jahrbuch Normative und institutionelle Grundfragen der Ökonomik, p 107 et seq

Weimann J, Hoffmann S (2003) Warum Messung und Bewertung biologischer Vielfalt nicht unabhängig voneinander möglich sind. In: Weimann J, Hoffmann A, Hoffmann S (eds) Messung und ökonomische Bewertung von Biodiversität – Mission impossible? Metropolis, Marburg, p 17 et seq

Weishaupt A, Ekardt F, Garske B, Stubenrauch J, Wieding J (2020) Land use, livestock, quantity governance, and economic instruments – sustainability beyond big livestock herds and fossil fuels. Sustainability 12:2053

Welzel C (2002) Fluchtpunkt Humanentwicklung. Über die Grundlagen der Demokratie und die Ursachen ihrer Ausbreitung. Westdeutscher Verlag, Wiesbaden

Wink R (2002) Generationengerechtigkeit im Zeitalter der Gentechnik. Evolutionär-institutionenökonomische Betrachtungen. Nomos-Verl.-Ges, Baden-Baden

Winter G (2001) Über Nutzen und Kosten der Effizienzregel im öffentlichen Recht. In: Gawel E (ed) Effizienz im Umweltrecht. Grundsatzfragen einer wirtschaftlichen Umweltnutzung aus rechts-, wirtschafts- und sozialwissenschaftlicher Sicht. Nomos-Verl.-Ges, Baden-Baden, p 97 et seq

Wurzer M (2014) Wirtschaftsethik von ihren Extremen her. Darstellung und Kritik der Ansätze von Karl Homann und Peter Ulrich. Königshausen & Neumann, Würzburg

Frictions on the Application Level: Costs and Benefits, Discounting, Uncertainty, Fact Base

3

Abstract

Although the above-mentioned well-known arguments against cost-benefit analysis are unconvincing, cost-benefit analysis fails because of certain unsolvable application problems. The frequently debated issues of dealing with risk, uncertainty and discounting of future concerns are by no means the only relevant aspects, but the very uncertainty of many cost-and-benefit effects mathematically precludes a monetarily exact quantification of decision situations. Even if leaving this aside, economists do not have a reliable way to represent the various cost-and-benefit effects that are not available in the form of market prices in a truly monetary way, despite numerous attempts. Inquiring about a hypothetical willingness to pay or supposedly observing the behavior of individuals (morals of the markets) leads to very incomplete, fictitious and, among other things, dispersed findings also due to very differently distributed abilities to pay. Empirically, the belief in infinite economic growth underlying cost-benefit analysis appears rather less robust. Thus, there are unsolvable application problems for the cost-benefit analysis. Those problems especially (but not exclusively) become visible in the context of climate protection, both in overall considerations of societal problems and in individual analyses, such as the construction of a coal-fired power plant.

3.1 Frictions of the Determination of Explicit and Implicit Preferences Respectively Costs and Benefits

Up to this point, it has been shown that certain classical objections against cost-benefit analysis are not convincing. Thus, the study turns to questions with respect to which resounding objections are indeed conceivable. First, inherent problems of

F. Ekardt, *Economic Evaluation, Cost-Benefit Analysis, Economic Ethics*, Environmental Humanities: Transformation, Governance, Ethics, Law 2, https://doi.org/10.1007/978-3-030-99284-2_3

application of cost-benefit analysis are considered (Chap. 3), then its compatibility with basic principles of liberal-democratic orders, which at the same time constitute legal orders, is discussed (Chap. 4), and finally the empiricist theoretical basis, which has already been considered historically, is ethically examined for its justification-theoretical tenability (Chap. 5).

Economists quantify all the concerns involved and then calculate what the optimal level of climate protection is, for example. In the process, everything that has a value for people, i.e., for which there is a preference, is to be translated into monetary units, right down to life and health.[1] There is no need for further balancing rules, the determined benefit or harm facts merge with the preferences and result in monetarized data at the end. One can already problematize whether the underlying normative preference theory is convincing as such. This is done later in Chap. 5. We first deal with the application problems of cost-benefit analysis, setting aside for the moment the fundamental concerns about the validity of the whole cost-benefit approach with its assumption of factual preferences to be monetized.

The thesis on the application problems is: There is already a lack of sufficiently precise data for benefits and costs that have a market price (example: damages to the tourism industry in Schleswig-Holstein due to climate change in the next 50 years). This is at least the case if, as in the case of climate change and biodiversity, the entire global economy is involved with an incalculable number of individual actions, highly complex effects, e.g., on ecosystems and environmental media such as soil and water (and their reaction to each other) and, moreover, time periods of more than 100 years, e.g., on ecosystems and environmental media such as soil and water (and their reaction among each other), complex social reaction attempts (as well as reactions to these reaction attempts) and, moreover, time periods of more than 100 years are involved. First, it is necessary to set this out in the following. Then, in the second part of the section, we analyze why the problems of costs and benefits that do not have a market price (such as human health) are magnified. Both types of problems have the effect that individual decisions cannot be answered meaningfully based on cost-benefit analyses. This is because these individual decisions raise exactly the same fundamental questions – what can one say about a coal-fired power plant, for example, if one does not include it in the overall consequences for climate, biodiversity, etc.?[2] Such questions are even more difficult to answer with a concrete assignment to exactly one power plant, since global effects are difficult to assign to a single industrial plant.

[1] Nordhaus 2008, p. 4; in detail on economic evaluation methods for human life Nestle 2012, p. 182 et seq.; Hammer 2011, p. 211 et seq.; Burtraw and Sterner 2009. Characterizing (and mostly critical) of preference-based, subjectivistic, naturalistic thinking also: von Kutschera 1982, p. 126 et seq. and passim.

[2] Economic voices in the literature (such as Hansjürgens and Lienhoop 2015, p. 62 et seq. and 107 et seq.; Hampicke 2001, p. 151 et seq.; Fromm 1997, p. 95 et seq. and 221 et seq.; Hanley and Barbier 2009, p. 315 et seq.; Nestle 2012, p. 77 et seq, 182 et seq. and 223 et seq.) miss the point if they assume that only large-scale questions and processes of change would be difficult to grasp.

The problem even for benefits and costs of processes with market prices arises in several respects, with the consequence of highly different (and thus not really usable) calculated costs and benefits in the respective economic studies[3]:

- There is first an obvious problem in dealing with risk and uncertainty. This will be considered separately in Sect. 3.2.
- From an ecological point of view, the comparison of different decision options (inherent in an economic evaluation) has limits, especially in a sub-area of environmental protection beyond climate protection, which is, however, strongly influenced by climate change, namely in nature conservation. This is because the comparison presupposes the substitutability of ecosystems, species or natural areas. In a strict sense, however, comparability does not exist, all the less so because nature is in a state of constant dynamic change.[4] Thus, cost-benefit analysis hides the complexity of nature, which makes it extremely difficult for sustainability issues.[5] This is also relevant for climate protection. For the consequences of climate change also include effects on ecosystems. Ignoring the lack of comparability and substitutability, at least when viewed strictly, also implies the concept known as weak sustainability, in which natural goods are regarded as goods that can be arbitrarily replaced by money. This is not true, however, because not only can ecosystems not be restored exactly in the form in which they were destroyed, but certain ecosystem services, once destroyed, cannot even be restored at all, including those that are of great or existential importance to humans.[6] It is true that biodiversity is ultimately to be regarded as a fundamentally renewable resource. However, even renewable resources can be overexploited and thus, beyond certain tipping points beyond which a collapse of the ecosystem concerned is to be feared, can be finite in a certain sense.[7]
- Another problem with the application of cost-benefit analysis, even with a mere orientation to costs and benefits with market prices, is that due to the diversity of environmental problems (even if one were to consider the problem of uncertainty to be non-existent) and the modern economy, immense amounts of data would have to be available. This is required to really calculate the costs and benefits of various options, for example, in climate protection. Such data is available for individual circumstances or areas, but not for nature as a whole. Moreover, dam-

---

[3] Susnjar 2010, p. 219 et seq.; Weimann and Hoffmann 2003, p. 17 et seq.; Spangenberg and Settele 2010, p. 327 et seq.; Baumgärtner 2003; Hoffmann et al. 2005, p. 25 et seq.; Ring and Schröter-Schlaack 2011, p. 16 et seq.; Rogall 2013, p. 213; Rogall 2015, p. 478 et seq.; see also Haensgen 2002, p. 10 et seq.; very incompletely (and without real justification optimistically) addressed in the allegedly synthesis by Brouwer et al., 2013, p. 56 et seq.

[4] On this pro toto Hanley and Barbier 2009, p. 138.

[5] Unmüßig 2014, p. 1; FoEE 2014, p. 3 et seq.; Hey 2012, p. 125 et seq.; Pissarskoi 2014, p. 90; Fromm 1997, p. 95 et seq.; SRU 2012, p. 35 et seq.; Hampicke 2001, p. 151 et seq.

[6] Unmüßig 2014, p. 1; FoEE 2014, p. 3 et seq.; Hanley and Barbier 2009, p. 323; Fromm 1997, p. 95 et seq.; ignoring this point: Pearce et al. 2006, p. 251 et seq.

[7] See on this Hansjürgens and Lienhoop 2015, p. 107 et seq.; Fromm 1997; Schlaudt 2016, p. 43 et seq.; Haensgen 2002, p. 11; Hampicke 2001, p. 151 et seq.

ages in the distant future depend on a large number of assumptions and are thus largely uncertain. Therefore, the calculation aimed at by the cost-benefit analysis lacks the necessary comprehensive factual basis. At best, one can work with avoidance costs (which can be recorded here and now), which, however, still pose further major problems (see below).[8]

As has already been indicated several times, even greater problems arise when costs and benefits have to be recorded without real market prices, i.e., when, according to the self-conception of cost-benefit analysis, it is a matter of willingness-to-pay analyses.[9] After all, manifold human damages resulting from failed environmental and nature conservation policies do not have market prices.[10] Economists therefore try to solve the problem of concerns without market value in ever new and often very differentiated ways, namely on the one hand by direct methods of inquired willingness to pay and on the other hand by analyses of indirectly revealed preferences (morality of the markets). A survey of revealed preferences, for example, wants to determine the value of natural areas based on the costs that people spend to be in the green or in its vicinity.[11] Alternatively, the hypothetical willingness of citizens to pay for the absence of hurricanes, for the continued existence of certain natural areas or, ultimately, for a higher probability of their own survival is surveyed directly. As described in Sect. 1.2, indirect methods address both costs and benefits, or both the provider and the demand side. As described, such approaches try to address opportunity costs, avoidance costs or recovery costs, or even travel costs.

Intensive economic research efforts (briefly documented in Sect. 1.2) have been directed at precisely comparing the performance of the individual approaches invoked in this way. The relevant debates also ask, for example, whether or not willingness-to-pay analyses lead to similar results in all cultural contexts; whether or not people do not show a willingness to pay simply because they consider something to be their right; or whether the aggregation of several study results can succeed (also when comparing, for example, the results of willingness-to-pay analyses on one environmental good and then on several environmental goods).[12] Little can and should be added to this here. Instead, we will address some rather fundamental

[8] Hampicke 2011; FoEE 2014, p. 3 et seq.; Unmüßig 2014, p. 1; largely confined to this problem is the otherwise methodologically rather carefree study by Brouwer et al., 2013, p. 60 et seq.

[9] On the following criticism Hausman 2012, p. 43 et seq.; Hansjürgens and Lienhoop 2015, passim; Pissarskoi and von Möllendorff 2013, p. 7 et seq.; Mathis 2009, p. 113 et seq.; Otsuka 2004; Meyer 2006, p. 136 et seq.; Hampicke 2011; Binswanger 2001, p. 67 et seq.; Pearce et al. 2006, p. 91 et seq. und 105 et seq.; Lienhoop and Hansjürgens 2010/2011, p. 255 et seq.; Hirschfeld 2012, p. 277 et seq.; UBA 2007, p. 93 et seq.

[10] See pars pro toto Unnerstall 1999, p. 180; Hausman 2012, p. 43 et seq.; admitted at Stern 2009, p. 92.

[11] For a detailed discussion of the economics of quantifying lost life years, see Pearce et al. 2006, p. 196 et seq.; Nestle 2012, p. 182 et seq.; Hammer 2011, p. 211 et seq.; Burtraw and Sterner 2009.

[12] See in detail on these points of view Endres and Holm-Müller 1998, p. 162 et seq.; Baumgärtner 2012, p. 82 et seq.; Hey 2012, p. 125 (128); Hansjürgens and Lienhoop 2015, p. 124 et seq.; Fromm 1997, p. 153 et seq.; Hausman 2012, p. 43 et seq.; Breunung 2001, p. 127 et seq.; Pearce et al. 2006, p. 282 and passim.

problems that make cost-benefit analysis seem unsuitable as a comprehensive decision-making and balancing construct, as it aims to be.

- First, even more than where market prices exist, the complexity of the preferences to be reconstructed for costs and benefits without market prices leads to unsolvable problems because of the complexity of climate change and nature in particular. It is not essential whether the preferences are to be elicited directly or reconstructed based on indirect methods. In any case, global climate change (as well as the development of ecosystems and biodiversity) affects such a variety of constellations that it cannot be mapped realistically. This becomes visible, for example, in the unfavorable consequences of climate change for biodiversity, which are extremely detailed. This problem is not the same as the issue of dealing with risk and uncertainty (Sect. 3.2), which further exacerbates the situation. But it is not only the question of what complexity can be handled at all and what can be known at all that arises. There is also the question of what the real people to be interviewed or observed really know, or if they are given extra information beforehand: what amount of information they can then process in real terms.[13] It is obvious, for example, that even a brief introduction of experts before the survey is hardly sufficient for someone to subsequently have such a professional overview of complex climate or biodiversity issues that he or she can express reflected preferences. Also, possible shifts to other environmental problems, to other sectors or other countries, for example, can hardly be captured in such a compressed way.[14]
- No matter how one tries to inquire about or observe willingness to pay: Ultimately, there is always a fictitious and therefore insufficiently informative element in determining what someone's own life or the absence of violent conflicts over resources, for example, is worth in monetary terms. This is especially true for explicit surveys. But even in the case of observation, one can merely observe another set of questions (such as payments for living in a green environment) whose connection to the willingness to pay for biodiversity or climate protection is only an indirect one. This is also critically noted in economics, without, however, drawing any further conclusions from it.[15] This is all the more true since it is well known from sociology and psychology that expressed attitudes of people often have little to do with their real behavior – and that the statements themselves often do not reflect the real attitudes, but can be massively distorted by lack of information, an adaptation to assumed intentions of the interviewer as well as to social desirability.[16] How little meaningful statements are can also be

[13] Also noted in economic writing by Hanley and Barbier 2009, p. 315 et seq.; Hansjürgens and Lienhoop 2015, passim; Hausman 2012, p. 43 et seq.

[14] See on this with further references Lange and Krull 2014, p. 166 et seq.; Ekardt 2019, § 6 D. IV.

[15] Hausman 2012, p. 43 et seq.; Hanley and Barbier 2009, p. 316; Hansjürgens and Lienhoop 2015, p. 50 et seq. und 66 et seq.; Fromm 1997, p. 153 et seq.; see also Nestle 2012, p. 182 et seq.

[16] The methodological problems of interdisciplinary behavioral research are discussed with further references in Ekardt 2019, Ch. 1.7.

seen from the fact that the willingness to pay for an environmental condition and the directly linked willingness to compensate differ significantly in surveys.[17]

- Observations of citizens' preferences (morality of markets) may be less fictitious on the whole than explicitly asking about willingness to pay; however, what is observed affects the preferences to be determined only indirectly or very indirectly. The value of lost income, for example, cannot possibly reflect the entire relevance of a death due to, say, climate change. Neither could the cost of treatment in the case of a climate change-related illness. Observations of the income for which people engage in a hazardous occupation, e.g., construction workers, cannot provide an overall picture either, partly because this would imply that the construction worker has voluntarily chosen his occupation and its precise risk (which he may not even know) by choosing between many alternatives. And why should people who are apparently willing to accept high risks be considered representative for the population? A shift to recovery costs does not really help either; on the one hand, this will often not reflect the full damage; and on the other hand, many things (e.g., deaths) cannot be recovered. Even countable points such as avoidance costs are not very meaningful in view of residual damage and multiple purposes for the avoidance expenditures – as well as the fact that in the end only a lower limit is marked in this way (e.g., in the representation of health damage). Still, the fact that security investments can be weighed against other desires in life is not disputed, since balancing as such is not criticized here (Sect. 2.2.3).[18] This is not to say, however, that it can usefully succeed in doing so by way of monetization and willingness-to-pay analyses, etc. To summarize: Millions of deaths and resource wars over water due to, say, climate change or a multitude of destroyed ecosystems would indeed also trigger costs in the monetary sense, some of which could perhaps be captured beyond market-priced services as restoration costs. However, it is obvious that the actually fatal aspect of such developments is only inadequately captured by referring to hospital costs, for example.
- Furthermore, the ability to pay naturally limits the willingness to pay, regardless of whether it is the willingness to pay that is asked for or the willingness to pay that is visible in action – with the consequence that the interests of a US billionaire would weigh massively much more than those of a Central African, because the latter can pay little or nothing. This is also realized by a climate economist like Nicholas Stern, contrary to the economic mainstream, and yet he too offers monetary values for non-market damages. If he then makes every person's life count equally in climate change, this is correct cum grano salis, but this happens without justification in the context of preference theory, which knows no normative standards.[19] Within the IPCC, there was some controversy around these issues and gradual (not very consistent) developments based on the fact that, in the context of the cost-benefit analysis, it seemed economically favorable, so to

[17] See on this Pearce et al. 2006, p. 155 et seq.; Hanley and Barbier 2009, p. 314 et seq.

[18] Ignored at Endres and Holm-Müller 1998, p. VII und 156 et seq.

[19] See on this Ekardt 2019, Ch. 3.8.

speak, that the victims of climate change would probably fall in large part among the less solvent population in the developing countries.[20]

- In addition (this is a well-known objection in economics): A cost-benefit analysis cannot correctly extrapolate people's factual preferences due to its empirically incorrect assumption of an eternally constant benefit of further income increases (linear utility function).[21] In short, the climate impact of coal-fired power plants had a different benefit in the past than it does today, and it has different benefits for different people today. Thus, one would have to survey preferences continuously and in a wide variety of situations, which seems almost impossible. At this point, one can also see that the survey approach leads to a necessary inconsistency with all the assumptions implicit in the cost-benefit analysis about the (normatively turned) homo oeconomicus, who always calculates rationally and consciously and would not tolerate any such inconsistencies.
- Another problem in the application of cost-benefit analysis is that it is applied very selectively. Neither in economics nor in politics is there any attempt to apply cost-benefit analysis comprehensively in all areas. If one continues to assume the validity of the approach for the time being, this does not seem very plausible. The quasi-plebiscitary intention based on factual preferences would suggest a comprehensive application, if at all. Thinking this through further, however, it becomes clear that cost-benefit analysis politically implies a program that could severely marginalize the economists who carry it out themselves: It ultimately suggests that liberal democracy should be comprehensively reorganized in a plebiscitary way for any individual decision as well as legislation, which ultimately implies computer-supported continuous voting. On closer inspection, however, it is precisely these plebiscitary tendencies that appear to be open to criticism, as will become clear later (Sect. 4.2).
- The mentioned points of criticism were already valid for the case that the cost-benefit analysis is comprehensively tried to be directed at all conceivable costs and benefits and that the different methods belonging to it are cumulatively applied. In reality, however, such a comprehensive coverage is usually not even attempted; rather, one finds in a very inconsistent way sometimes this, sometimes that access to costs and benefits with regard to the respectively used direct and indirect methods of coverage.[22] However, this makes the data collection even more incomplete and even less meaningful than it already is due to the application problems described above.

The above criticisms do not change when ethical considerations are officially added to the cost-benefit analysis (as in the IPCC's Fifth Assessment Report).[23] In the sense of the empiricist epistemological model of economics, this again means

[20] For a summary see Jamieson 2014, p. 115 et seq.; furthermore, on the ability to pay: Fromm 1997, p. 22.

[21] On the mathematical details: Breyer and Kolmar 2014, p. 65 et seq.

[22] Mentioned in Hanley and Barbier 2009, p. 308; Arrow et al., 1997, p. 196 et seq.

[23] See IPCC 2014, WG III, Ch. 2.6. Even to the extent that voices of philosophical discourse are explicitly noted there, they are such voices that themselves argue in a more or less strongly preference-based manner, such as Amartya Sen (who is primarily an economist) or utilitarian thinkers.

something empirical (and nothing normative), e.g., the determination of the factual altruistic preferences of people. Ultimately, this does not add anything new at all, because even in the classical determination of preferences, individuals are not prevented from articulating altruistic preferences.[24] This example underlines once again that in economics it is mostly not correctly grasped that the normative system of ethics does not coincide with an empirical moral sociology. Ergo, all the application problems mentioned also affect the supposedly determined ethical preferences.

The aforementioned criticisms are repeated again, mutatis mutandis, when irreplaceable goods are to be excluded from the cost-benefit balancing in more recent cost-benefit approaches.[25] In an empiricist theoretical framework that knows no normative rightness or wrongness beyond factual preferences, who is to determine which goods are "irreplaceable"? And what does "irreplaceable" mean anyway; an old vase may seem irreplaceable to someone, but does this mean that one is obliged to preserve it without any balancing, i.e., also at arbitrarily high costs and for all time? Or what about a stable global climate or biodiversity; at least if humanity prefers short-term pleasure and profit-seeking to long-term survival, there is little to be said for irreplaceability here. And a recent observation of the morality of markets as revealing the factual preferences of consumers, corporations, and perhaps politicians suggests that those preferences tend not to swing in favor of protecting climate and biodiversity. Talk of irreplaceable goods therefore presupposes the assumption of an ethical default that purports such irreplaceability. However, a cost-benefit analyst can hardly assume the existence of such a rule, since the theoretical basis of his cost-benefit analysis is precisely the idea that there is no meaningful talk about normative issues beyond empiristic cost-benefit analysis.

The application problems of cost-benefit analysis cannot be remedied either if one questions the basic intention of cost-benefit analysis even further and asks for "satisfactory" trade-off results instead of "optimal" ones. First, this does not change any of the points raised. Second, this would require a normative criterion that shows which condition is to be classified as satisfactory.[26] This could be determined in the basic intention of the cost-benefit analysis again only on the basis of factual preferences, however, without any problem disappearing thereby.

It is astonishing that in economics (as can be seen from the citations) many of the problems considered in this section are known and yet cost-benefit analysis continues to be resolutely propagated.[27] An interesting approach, on the other hand, might be to conceive of cost-benefit analysis from the outset not as primary, but merely as complementary to ethics and law, limited to goods with market prices – where they

[24] This is also explicitly demanded, for example, by Lienhoop and Hansjürgens 2011, p. 229 et seq.; Stern 2009 also contains many such formulations.

[25] Skeptical also Anderson et al. 2015.

[26] Ignored at Simon 1993; too optimistic also Susnjar 2010, p. 226 et seq.

[27] Visible at Endres and Holm-Müller 1998, p. 69; Hansjürgens and Lienhoop 2015, p. 11 et seq.; aptly: Schlaudt 2016, p. 138 et seq.; Hausman 2012, p. 43 et seq.

can be readily captured.[28] This option and its reconciliation with legal or ethical balancing will be considered in the final section (Sect. 6.2).

3.2 Problems of Economic Dealing with Risk and Uncertainty

A particularly large and structurally unsolvable problem in the quantification of cost-benefit analysis has only been hinted at so far and will now be discussed separately. This problem consists in the fact that generally uncertain future events cannot be assigned a numerical value per se, if not even the probability of occurrence and the possible extent of damage can be determined exactly. This is true even for goods that have a market price. If, for example, we consider how climate change will affect harvests, grain prices, flood damage, the frequency of storms, the frequency of civil wars and wars, certain diseases, and above all (separately) the course and the amount of the costs of these factors, we are obviously faced with manifold uncertainties.[29]

Because of this aspect alone, economic statements on such complex events as global climate change or biodiversity loss and the infinite variety of economic cause and effect factors associated with them can only be based on ultimately subjectively determined assumptions. Because if one does not know how certain a certain event will occur and with what exact consequences, the preferences asked about it also become very vague. Consequently, it is stated even in economics that future uncertain events cannot actually be integrated into precise cost calculations.[30] Of course, this is especially true when, as in the case of climate change, the entire global economy is involved, with an incalculable number of individual actions, highly complex effects, e.g., on ecosystems and environmental media such as soil and water, complex social response attempts (and reactions to these response attempts), and, moreover, time periods of several centuries. The same results from the fact that the interactions within ecosystems, between different ecosystems, and the effects of changes in ecosystems or in bio-diversity on the provision of ecosystem services are highly complex and in some cases difficult to understand in detail.[31]

Ultimately, however, this also applies to any other social decision, at least cum grano salis, since complexity is an inherent characteristic of collective decision-making situations. The problem cannot be solved by thresholds and other artificial

---

[28] Cum grano salis, such an approach can then also indicate which states are to be considered satisfactory; to that mental connection Susnjar 2010, p. 226 et seq.

[29] For a list of factors see Stern 2007, p. 146.

[30] Pissarskoi and von Möllendorff 2013, p. 7 et seq.; Wink 2002, p. 146 et seq.; Haensgen 2002, p. 11; Wätzold 2000, p. 299 et seq.; UBA 2007, p. 23 et seq.; Sunstein 2005, p. 351 et seq.; Künzler 2012; Hampicke 2001, p. 151 et seq.; Meggle 2010, p. 125; Löhr 2009, p. 40 et seq. und 62 et seq.; Fromm 1997, p. 219; Hansjürgens and Lienhoop 2015, p. 109; Hanley and Barbier 2009, p. 315 et seq.; in detail on economics and uncertainty Nestle 2012, p. 30 et seq.

[31] TEEB 2010; see also Ring and Schröter-Schlaack 2011, p. 16 et seq. and 186 et seq.; Ring and Schröter-Schlaack 2011, p. 15 (16).

definitions, as they are proposed.[32] In this way, cost-benefit analysis would once again violate its own theoretical foundations: to orient itself solely on the factual preferences of people. Even limiting the cost-benefit analysis to small-scale cases such as the effects of a coal-fired power plant does not solve the problem; as already mentioned, this repeats the question of the various complex climate and biodiversity effects, which is even more difficult to answer with a concrete assignment to exactly one power plant. Also, the attempt to simply deal with uncertainty as such by observing, for example, people performing hazardous activities, has already been indirectly criticized above, because it would be assumed that the construction worker has voluntarily chosen his activity and its exact risk (which he may not even know). And why should people who are apparently willing to accept high risks be considered a representative cross-section of the population?[33]

One option, of course, remains: one could dispense with any indirect survey and with any use of market prices as a basis, and instead simply ask for preferences directly for any political or administrative decision, with the idea that the respondents do not know the probabilities of occurrence (simply because they are not known to anyone). However, this would no longer be a cost-benefit analysis, but simply a qualitatively comprehensive plebiscitary decision-making system (for a critical discussion of the plebiscitary idea, see Sect. 4.2).

3.3 Problems of the Debate on Discounting

Another application problem of cost-benefit analyses concerns how to deal with the fact that certain costs and benefits occur only at some distance in time. Even if these costs and benefits are certain, it is almost unanimously assumed among economists that such temporally distant effects should matter less than current effects-and that an empirically assumed individual discounting propensity can simply be applied to society as a whole. Like the uncertainty debate, this point has already been widely discussed in the literature (including in ethics), so that only a brief comment is made on it here.[34]

The idea of discounting is plausible, at least in the first moment, if the victim of the damage is the same person today and in ten years. This is especially true within the basic principles of cost-benefit analysis, which (in itself) is based on factual

[32] On the different approaches see UBA 2007, p. 23 et seq.

[33] Furthermore, it is ignored that even from figures that reflect facts, very different conclusions can often be drawn. An interesting example of this is the (real or supposed) deterrent effect of the death penalty in Kirchgässner 2011a, b, p. 18 et seq.

[34] Critical of discounting Unnerstall 1999, p. 320 et seq.; Hampicke 2011; Jamieson 2014, p. 115 et seq.; Pissarskoi and von Möllendorff 2013, p. 7 et seq.; Pissarskoi 2014, p. 79 et seq. ; open Birnbacher 1988; Hanley and Barbier 2009, p. 142 et seq.; Fromm 1997, p. 236 et seq.; Goulder and Stavins 2002, p. 673 et seq.; Pearce et al. 2006, p. 190 et seq.; Mathis 2011, where discounting against the yardstick of alternative investments is considered permissible (a follow-up question would be the growth orientation therein); precisely not different (though this is often overlooked) Ramsey 1928.

preferences. And discounting of own (!) future costs and benefits is mostly practiced by people from a purely factual point of view.[35] Nor need one be bothered by the fact that descending discount rates could not be meaningfully formed at some point given the infinity of the future, since the human possibility of causing costs and benefits does not extend arbitrarily far into the future anyway.[36] Nevertheless, the discounting approach of cost-benefit analysis for all costs and benefits meets with rigorous objections:

- First, in the systematics of the preference approach, the problematic point (important in sustainability issues) is that the preferences of those affected, especially future generations, cannot be meaningfully captured by the approach. Ultimately, of course, this is a problem that, because of its inevitable inherence in economic evaluation, affects the fundamentals and not just the application of cost-benefit analysis. It will therefore be addressed in more detail later (Sect. 5.1).
- Furthermore, the supra-personal character of discounting is problematic, i.e., that not only everyone should discount their own costs and benefits. The point that interpersonal balancing of interests can and must take place (Sect. 2.2.3) and that this balancing does not in principle have to dispense with quantifying elements (Sect. 2.2.4) has already been discussed. This does not mean, however, that this balancing can take place like a normal addition process which is supposedly not in need of further justification. Why should the damage of a Bangladeshi in 50 years be per se less important than my damage today? The preference approach implies that everyone is responsible for himself, i.e., that he does not pass on the consequences of his own actions to others. Moreover, there is much to be said for the existence of protection guarantees also in favor of people in the future, which preclude ignoring them per se.[37] Discounting cannot be justified by a general expectation of perpetual growth either, whether for those already alive today or for future generations, as becomes apparent with the insight into the limits of growth (Sect. 3.4). Furthermore, discounting cannot be justified via compensations of future generations, because such a discounting is not possible in a meaningful way, apart from the fact that it would have to be established first.[38] One could say: Future people cannot yet express preferences, so they are not interesting. This would be, as already mentioned, the statement directly implied in the preference theory. In this case, however, one would not have to discount but to declare all damages that affect someone who is not yet alive today as irrelevant, otherwise the preference theory would become self-contradictory (apart from this, it will be shown later that such a fading out of the future is rather unconvincing: Sect. 5.1).

---

[35] Goulder and Stavins 2002, p. 673 et seq.; Hanley and Barbier 2009, p. 163; Pearce et al. 2006, p. 190 et seq.

[36] Mentioned at Moellendorf 2014, p. 90 et seq.; Pearce et al. 2006, p. 190 et seq.

[37] See in more detail Ekardt 2019, Ch. 3.3; Unnerstall 1999.

[38] See in more detail Fromm 1997, p. 247 et seq.

- Even with respect to people living today, discounting purely for the sake of the passage of time is inconsistent if it is not itself raised directly or indirectly as a preference, but is simply set. Why should an economic theorist, based on preference theory, be allowed to tell me whether I have a present preference and should not care about the future? Concerning the next step of the supposedly conceivable observation of possible discounting preferences on the basis of a morality of the markets some thoroughgoing points of criticism have already been formulated above. This approach, too, does not allow the derivation of a discount rate. There are simply no observable market or interest rate developments that would say anything at all about which factual preferences exist with respect to damages over several centuries – and with irreversible character. Similarly, no observable preferences exist to the effect that people are willing to discount across interpersonal boundaries. And even for people themselves, discounting usually proceeds much more inconsistently than assumed.[39]
- And even if all this were disregarded, discounting would in any case only be conceivable if the damage to be discounted can actually be represented in monetary terms despite the above-mentioned criticism. And it tends not to be. The reason is that discounting carries around as a rucksack all the application problems of cost-benefit analysis mentioned above, including the unclear handling of uncertainty.

However, the famous climate economist Nicholas Stern puts forward an argument for discounting that is at least debatable: the uncertain probability of future damage events. The fact that it reduces the weight of normative concerns in balancing in the abstract has been noted elsewhere.[40] In this respect, too, the problem remains for a quantifying balancing system such as cost-benefit analysis as to whether this can be expressed mathematically. In any case, if no mathematical probability can be given for the future events, an allegedly clear discount rate is ultimately arbitrary. In the end, even the conceivable change in opportunity costs does not change this, because certain costs and benefits do not necessarily have to become monetarily more favorable in the future.[41] They may also become more expensive, for example, due to increased scarcity.[42]

[39] Elaborated in more detail at Hampicke 2011.

[40] See Ekardt 2019, Ch. 3.7; Ekardt 2021, § 5 C. II. 2.

[41] Mathis 2011, p. 165 et seq.; Hammer 2011, p. 211 et seq.

[42] Mentioned in Jamieson 2014, p. 115 et seq.

3.4 Frictions of the Empirical Foundations: Especially on the Technology and Growth Orientation of Cost-Benefit Analysis

Another problem with the application of cost-benefit analyses is that they are often based on overly optimistic facts with regard to ecological problems when they attempt to forecast future damage from climate change (even if this damage can be expressed in market prices). For example, the need to reduce greenhouse gas emissions to zero within a few years, as described above, is not taken as a basis. The fact that cost-benefit analyses use outdated factual statements, e.g., on climate change, could still be dismissed as a flaw in practical implementation that could in principle be easily remedied.[43] However, the cost-benefit analysis appears to be so heavily influenced by such assumptions that the error cannot simply be corrected. This is true, as will now be shown, at least with respect to economic growth.

The identification of a material increase as welcome or efficient, as it is inherent in cost-benefit analysis, already implies per se an orientation towards growth. At the same time, economists assume that such a constant growth corresponds to the preferences of individuals in the sense that the constant material, i.e., monetary, increase makes them happier (whether this would be conducive to human happiness and to what extent this could be judged normatively and empirically at all was a topic elsewhere, but shall not be of interest again here).[44] Consequently, economics today almost unanimously assume the possibility and desirability of endless economic growth, based on infinitely possible technical innovations whose market introduction could overcompensate for the lost growth effects of old technologies (such as fossil fuels).[45] The idea of an eternally growing material cake can also be understood as a justification of the inequality in voting power (or in the ability to pay) in cost-benefit analysis. Without getting into the complex philosophical debate on equality here, there is in any case a multiple major problem with the empirical presuppositions of eternal growth. In abbreviation of a likewise large debate, only individual central points of view are to be referred to here.

If, for example, the Paris climate targets were achievable with a purely technical strategy, it can already be controversially discussed whether the increase in renewable energies and energy efficiency can actually fully compensate or even overcompensate for the loss of fossil fuels in terms of economic growth.[46] However, this becomes even more doubtful when one sees that the purely technical path will prob-

[43] See as an example Nordhaus 2008, p. 5 et seq. und 123 et seq.; Paqué 2010, p. 102 et seq.

[44] See with further references Ekardt 2019, Ch. 2.6 and 3.4; more optimistic regarding the empirical happiness potentials of degrowth (and the normative answerability of the question of right happiness) with different nuances Hosang et al. 2005; Jensen and Scheub 2015; Paech 2014.

[45] Aptly at Gorz 1989, p. 16 and passim; Schlaudt 2016, p. 137 et seq.; as an example Paqué 2010, p. 102 et seq.; Stern 2007.

[46] See on the following Shindell 2015, p. 313 et seq.; Machol and Rizk 2013, p. 75 et seq.; Fücks 2013, p. 105 et seq.; Jakob and Edenhofer 2014, p. 447 et seq.; Ismer 2014, p. 65 et seq.; Paqué 2010, p. 102 et seq.; Ekardt 2019, Ch. 1.4; Stern 2007.

ably not suffice. The extent of the problem, for example, in the case of climate change, points in this direction. Measured by the speed of innovation known so far, it seems only moderately probable that a change to renewable energies and energy efficiency alone will lead to a complete phase-out of fossil fuels before 2030 or 2040, as would be necessary for the emissions budget to achieve a global 1.8 or 1.5-degree limit in accordance with Art. 2 of the Paris Agreement as an international climate protection target (Sect. 1.3). As mentioned above, there are further aspects pointing in this direction.

The need for frugality shows that the above-mentioned challenge to the growth paradigm actually exists. Growth and its safeguarding is seen by many as the central political-social goal, in a country and worldwide. It is expected to lead to greater well-being, stable welfare states, increased human happiness, greater freedom, the creation of jobs and much more. Also, in developing and emerging countries, the fight against often dramatic poverty in the coming decades implies some kind of economic growth. At the same time, however, economic growth is a key driver of climate and resource problems, in that growth also increases the consumption of fossil fuels, despite all the possibilities of green growth. This is precisely what emerging economies are emulating. If, as seen, frugality must be a key part of the sustainability shift, then less will be sold (such as fewer vacation flights). This is precisely what could mean, if it assumes a significant scale, an unplanned transition to a post-growth society, i.e., a society that has to manage permanently without growth or even adapt to processes of shrinkage.[47]

Certainly, individual business ideas can be developed from the concept of frugality that have to do with starting points such as sharing, regional, slow, service orientation or straight educational measures and courses. Individual companies could grow with this. All in all, however, genuine frugality, if it achieves its ecological goals and is not cancelled out by rebound effects or the shifting of problems to other countries, other sectors or other environmental problems, would probably consist precisely of all of us buying less. This will most likely mean that the current growth society will not be able to continue as it is. This foreseeable finite nature of growth is a major problem, since modern societies have been dependent on economic growth to a large extent up to now. This requires solutions for areas of society that have so far been partly dependent on growth, such as the labor market, public debt, the tax system, pension insurance and the banking system. And this involves solutions both for the transition and for the imagined target state. These problems may be solvable, but they are significant.[48]

In any case, with the end of the growth era, cost-benefit analysis latently loses a central empirical basis. However, this is not an argument against it with the same clarity as the other aspects discussed here. For if it were possible (which, however, is not the case, see above) to actually represent all costs and benefits – even those

[47] Jackson 2016; Ward et al., 2016; Paech 2014; Schulz and Bailey 2014, p. 277 et seq.; Scheidler 2015, p. 205 et seq.; Stengel 2011, p. 163 et seq.; Ekardt 2019, Ch. 1.4; for a more optimistic perspective see Rogall 2012, p. 163 et seq.

[48] Calculated, for example, for Canada by Victor and Rosenbluth 2007, p. 492 et seq.

without market prices – in monetary terms, then the cost-benefit analysis could still provide indications as to how the shrinking process can proceed as favorably as possible.

Summarizing the present chapter, it can be stated that this is not the only application problem of cost-benefit analysis. In fact, the application of cost-benefit analysis fails due to the complexity of the issues, the lack of market prices for many issues and other factors. The frequently debated issues of dealing with risk, uncertainty and discounting are by no means the only relevant aspects. And application problems in general are also only part of the frictions of cost-benefit analysis, as will be shown below.

Repetition Questions

1. What problems arise in cost-benefit analysis in calculations with real preferences? (Sect. 3.1)
2. Which problems arise in cost-benefit analysis in calculations with hypothetical preferences? (Sect. 3.1)
3. Can cost-benefit analysis solve the problem of dealing with uncertainty? (Sect. 3.2)
4. Is discounting of future preferences justified? (Sect. 3.3)
5. Which other problems arise with the factual basis of cost-benefit analysis, such as climate change? (Sect. 3.4)
6. Where does economic growth as a paradigm reach its limits, what is meant by post-growth, and to what extent does post-growth in turn create problems? (Sect. 3.4)

References[49]

Anderson M, Teisl M, Noblet C, Klein S (2015) The incompatibility of benefit–cost analysis with sustainability science. Sustain Sci 10(1):33–41. https://doi.org/10.1007/s11625-014-0266-4

Arrow K et al (1997) Is there a role for benefit-cost analysis in environmental, health and safety regulation? Environ Dev Econ 272:196 et seq

Baumgärtner S (2003) Warum Messung und Bewertung biologischer Vielfalt nicht unabhängig voneinander möglich ist. In: Weimann J, Hoffmann A, Hoffmann S (eds) Messung und ökonomische Bewertung von Biodiversität – Mission impossible? Metropolis, Marburg, p 43 et seq

Baumgärtner S (2012) Normative Begründung der Nachhaltigkeitsökonomik. In: Studierendeninitiative Greening the University (ed) Wissenschaft für nachhaltige Entwicklung! Multiperspektivische Beiträge zu einer verantwortungsbewussten Wissenschaft. Metropolis-Verlag, Tübingen, p 273 et seq

Binswanger M (2001) Technological progress and sustainable development: what about the rebound effect? Ecol Econ 36:119 et seq

[49] In accordance with legal practice, parliamentary, governmental and EU Commission documents as well as laws and judgments are not listed in the bibliography, as they can be found unequivocally based on the reference given in the continuous text or via the general search engines. The last access date for all Internet sources is 31.10.2021.

Birnbacher D (1988) Verantwortung für zukünftige Generationen. Stuttgart
Breunung L (2001) Umweltschutz unter dem Optimalitäts- und Effizienzpostulat. Die wohlfahrtsökonomische Konzeption in soziologischer Sicht. In: Gawel E (ed) Effizienz im Umweltrecht. Grundsatzfragen einer wirtschaftlichen Umweltnutzung aus rechts-, wirtschafts- und sozialwissenschaftlicher Sicht. Nomos, Baden-Baden, p 127 et seq
Breyer F, Kolmar M (2014) Grundlagen der Wirtschaftspolitik, 4th edn. Mohr Siebeck, Tübingen
Brouwer R et al (2013) A synthesis of approaches to assess and value ecosystem services in the EU in the context of TEEB. http://ec.europa.eu/environment/nature/biodiversity/economics/pdf/EU%20Valuation.pdf
Burtraw D, Sterner T (2009) Climate Change Abatement. http://www.rff.org/Publications/WPC/Pages/09_04_06_Climate_Change_Abatement.aspx
Ekardt F (2019) Sustainability. Transformation, governance, ethics, law. Springer, Dordrecht
Ekardt F (2021) Theorie der Nachhaltigkeit. Ethische, rechtliche, politische und transformative Zugänge – am Beispiel von Klimawandel, Ressourcenknappheit und Welthandel, 4th edn. (= 3rd ed. of the new edition). Baden-Baden
Endres A, Holm-Müller K (1998) Die Bewertung von Umweltschäden. Theorie und Praxis soziooökonomischer Verfahren. Kohlhammer, Stuttgart et al
Friends of the Earth Europe (FoEE) (2014) Nature is not for sale. The dangers of commodifying our natural world. http://www.foeeurope.org/sites/default/files/foee_position_nature_is_not_for_sale.pdf
Fromm O (1997) Möglichkeiten und Grenzen einer ökonomischen Bewertung des Ökosystems Boden. P. Lang, Frankfurt a.M
Fücks R (2013) Intelligent wachsen. Die grüne Revolution. Munich
Gorz A (1989) Kritik der ökonomischen Vernunft. Sinnfragen am Ende der Arbeitsgesellschaft. Rotbuch, Berlin
Goulder L, Stavins R (2002) Discounting. An eye on the future. Nature 419:673 et seq
Haensgen T (2002) Das Kyoto-Protokoll: Eine ökonomische Analyse unter besonderer Berücksichtigung der flexiblen Mechanismen, Working paper. Bamberg
Hammer B (2011) Valuing the invaluable? Valuation of human life in cost-efficient assessments of regulatory interventions. In: Mathis K (ed) Efficiency, sustainability, and justice to future generations. Springer, Berlin, p 211 et seq
Hampicke U (2001) Grenzen der monetären Bewertung. Kosten-Nutzen-Analyse und globales Klima. Jahrbuch Ökologische Ökonomik:151 et seq
Hampicke U (2011) Climate change economics and discounted utilitarianism. Ecol Econ 72:45 et seq
Hanley N, Barbier E (2009) Pricing nature. Cost-benefit analysis and environmental policy. Elgar, Cheltenham
Hansjürgens B, Lienhoop N (2015) Was uns die Natur wert ist. Potenziale ökonomischer Bewertung. Metropolis-Verlag, Marburg
Hausman J (2012) Contingent valuation – from dubious to hopeless. J Econ Perspect 26:43 et seq
Hey C (2012) Wege aus dem Wachstumsdilemma. Kritische Anmerkungen zu einer aktuellen Debatte. Zeitschrift für Umweltpolitik und Umweltrecht 35:125 et seq
Hirschfeld J (2012) Kosten und Nutzen der Anpassung an den Klimawandel – ein Verteilungsproblem. In: Ekardt F (ed) Klimagerechtigkeit. Ethische, rechtliche, ökonomische und transdisziplinäre Zugänge. Metropolis-Verlag, Marburg, p 277 et seq
Hoffmann A, Hoffmann S, Weimann J (2005) Irrfahrt Biodiversität. Eine kritische Sicht auf europäische Biodiversitätspolitik. Metropolis-Verl, Marburg
Hosang M, Fraenzle S, Markert B (2005) Die emotionale Matrix. Grundlagen für gesellschaftlichen Wandel und nachhaltige Innovation. Oekom-Verl, Munich
IPCC (2014) Fifth Assessment Report. www.ipcc.int
Ismer R (2014) Klimaschutz als Rechtsproblem. Steuerung durch Preisinstrumente vor dem Hintergrund einer parallelen Evolution von Klimaschutzregimes verschiedener Staaten. Mohr Siebeck, Tübingen
Jackson T (2016) Prosperity without growth, 2nd edn. London

Jakob M, Edenhofer O (2014) Growth, degrowth, and the commons. Oxf Rev Econ Policy:447 et seq
Jamieson D (2014) Reason in a dark time. Why the struggle against climate change failed – and what it means for our future. Oxford University Press, Oxford
Jensen A, Scheub U (2015) Glücksökonomie. Munich
Kirchgässner G (2011a) Der Ökonom als Berater – Objektivität, Ideologie und Eigeninteresse. Jahrbuch normative und institutionelle Grundfragen der Ökonomik:219 et seq
Kirchgässner G (2011b) Econometric estimates of deterrence of the death penalty – facts or ideology? CESifo Working Paper No. 3443. University of St. Gallen, Munich
Künzler A (2012) The jurisprudence of welfare maximization. Embracing complexity and uncertainty in the environmental, health and safety context. Archiv für Rechts- und Sozialphilosophie 98:1 et seq
Lange C, Krull D (2014) Infrastrukturelle Wasserdienstleistungen. Zur Funktion kostendeckender Wasserpreise nach Art. 9 der EG-Wasserrahmenrichtlinie und ihre mögliche Anwendung auf den Aufstau von Gewässern als infrastrukturelle Wasserdienstleistung für Binnenschifffahrt und Wasserkraft. Metropolis-Verl, Marburg
Lienhoop N, Hansjürgens B (2010/2011) Vom Nutzen der ökonomischen Bewertung in der Umweltpolitik. GAIA:255 et seq. und 229 et seq
Löhr D (2009) Die Plünderung der Erde. Metropolis-Verl, Marburg
Machol B, Rizk S (2013) Economic value of U.S. fossil fuel electricity health impacts. Environ Int 52:75 et seq
Mathis K (2009) Efficiency instead of justice? Searching for the philosophical foundations of the economic analysis of law. Springer, Berlin
Mathis K (2011) Consequentialism in law. In: Mathis K (ed) Efficiency, sustainability, and justice to future generations. Springer, Berlin, p 3 et seq
Meggle G (2010) Handlungstheoretische Semantik. Berlin et al, De Gruyter
Meyer K (2006) How to be consistent without saving the greater number. Philos Public Aff 34:136 et seq
Moellendorf D (2014) The moral challenge of dangerous climate change. Values, poverty, and policy. Cambridge University Press, Cambridge, MA
Nestle I (2012) The costs of climate change in the agricultural sector. A comparison of two calculation approaches, Dissertation, Flensburg
Nordhaus W (2008) A question of balance. Weighing the options on global warming policies. Yale University Press, New Haven et al
Otsuka M (2004) Skepticism about saving the greater number. Philos Public Aff 32:413 et seq
Paech N (2014) Liberation from excess. Munich
Paqué K-H (2010) Wachstum! Die Zukunft des globalen Kapitalismus. Hanser, Carl, Munich
Pearce D, Atkinson G, Mourato S (2006) Cost-benefit analysis and the environment. Recent development. Paris
Pissarskoi E (2014) Gesellschaftliche Wohlfahrt und Klimawandel. Umgang mit normativen Annahmen und Ungewissheiten bei der klimaökonomischen Politikberatung. Oekom-Verl, Munich
Pissarskoi E, von Möllendorff C (2013) Ökonomie der Anpassung an den Klimawandel. Literaturauswertung: Bewertung von Klimafolgen und Kosten-Nutzen-Analysen von Anpassungsstrategien in Deutschland. Berlin
Ramsey FP (1928) A mathematical theory of saving. Econ J:543 et seq
Ring I, Schröter-Schlaack C (eds) (2011) Instrument mixes for biodiversity policies. POLICYMIX Report, Issue No. 2/2011, Leipzig
Rogall H (2012) Nachhaltige Ökonomie. Ökonomische Theorie und Praxis einer Nachhaltigen Entwicklung, 2nd edn. Marburg (3rd ed. angekündigt für 2018)
Rogall H (2013) Volkswirtschaftslehre für Sozialwissenschaftler. Einführung in eine zukunftsfähige Wirtschaftslehre, 2nd edn. Springer, Wiesbaden
Rogall H (2015) Grundlagen einer nachhaltigen Wirtschaftslehre. Volkswirtschaftslehre für die Studierenden des 21. Jahrhunderts, 2nd edn. Marburg
Sachverständigenrat für Umweltfragen (2012) Umweltgutachten 2012. Berlin

Scheidler F (2015) Das Ende der Megamaschine. Geschichte einer scheiternden Zivilisation. Promedia, Wien
Schlaudt O (2016) Wirtschaft im Kontext. Eine Einführung in die Philosophie der Wirtschaftswissenschaften in Zeiten des Umbruchs. Vittorio Klostermann GmbH, Frankfurt a.M
Schulz C, Bailey I (2014) The green economy and post-growth regimes – opportunities and challenges for economic geography. Geogr Ann 96:277 et seq
Shindell D (2015) The social cost of atmospheric release. Clim Chang 130:313 et seq
Simon H (1993) Homo rationalis. Frankfurt a.M
Spangenberg J, Settele J (2010) Precisely incorrect? Monetising the value of ecosystem services. Ecol Complex:327 et seq
Stengel O (2011) Suffizienz. Die Konsumgesellschaft in der ökologischen Krise. oekom verlag, Munich
Stern N (2007) Stern review. The economics of climate change. HM Treasury, Cambridge
Stern N (2009) A blueprint for a safer planet. How to manage climate change and create a new era of progress and prosperity. Bodley Head, Cambridge
Sunstein C (2005) Law of fear. Beyond the precautionary principle. Cambridge University Press
Susnjar D (2010) Proportionality, fundamental rights, and balance of powers. Martinus Nijhoff Publishers, Leiden
TEEB (2010) The economics of ecosystems and biodiversity: the ecological and economic foundations
Umweltbundesamt (UBA) (2007) Ökonomische Bewertung von Umweltschäden. Methodenkonvention zur Schätzung externer Umweltkosten. Dessau
Unmüßig B (2014) Monetizing nature – taking precaution on a slippery slope. http://us.boell.org/2014/08/26/monetizing-nature-taking-precaution-slippery-slope
Unnerstall H (1999) Rechte zukünftiger Generationen. Königshausen und Neumann, Würzburg
Victor P, Rosenbluth G (2007) Managing without growth. Ecol Econ 61:492 et seq
von Kutschera F (1982) Grundlagen der Ethik. De Gruyter, Berlin
Ward J et al (2016) Is decoupling GDP growth from environmental impact possible? PLOS ONE. https://doi.org/10.1371/journal.pone.0164733
Wätzold F (2000) Efficiency and applicability of economic concepts dealing with environmental risk and ignorance. Ecol Econ 2(33):299–311
Weimann J, Hoffmann S (2003) Warum Messung und Bewertung biologischer Vielfalt nicht unabhängig voneinander möglich sind. In: Weimann J, Hoffmann A, Hoffmann S (eds) Messung und ökonomische Bewertung von Biodiversität – Mission impossible? Metropolis, Marburg, p 17 et seq
Wink R (2002) Generationengerechtigkeit im Zeitalter der Gentechnik. Evolutionär-institutionenökonomische Betrachtungen. Nomos-Verl.-Ges, Baden-Baden

4 Collision of the Cost-Benefit Analysis with Liberal-Democratic Basic Principles and the Claim of Validity of Law

Abstract

There are collisions between cost-benefit analysis and liberal-democratic constitutional law, the principles of which are also fundamental ethical principles. This concerns both the rights to freedom and its elementary preconditions as well as the model of democracy and respect for the rule of law. The weight of the fundamental rights and the balancing rules applicable in the case of conflicting spheres of freedom in liberal democracies – which, in short, form the basis of the legal decision-making scheme (then further concretized in law in many details) – do not depend on how much someone can pay. Freedom in a liberal democracy is not only the freedom of solvent consumers with their preferences, but also the freedom of those who are hardly present in the free market due to a lack of purchasing power and who would suffer massively from, for example, the consequences of climate change and damaged ecosystems. Furthermore, for good reasons, decision-making in liberal democracies is representative-democratic with only isolated plebiscitary (and participatory) additions. This is contrary to the plebiscitary intentions of the cost-benefit analysis. Also, for good reasons, confirmed by centuries of historical experience, modern law claims a monopoly of the decision of normative questions for itself, which only knows an opening in favor of other decision-making procedures, such as the free play of forces, where this is permitted by law. A competing normativity to the law on the part of the cost-benefit analysis is thus excluded.

F. Ekardt, *Economic Evaluation, Cost-Benefit Analysis, Economic Ethics*, Environmental Humanities: Transformation, Governance, Ethics, Law 2,
https://doi.org/10.1007/978-3-030-99284-2_4

4.1 Common Economic Views on Law – And Legal Views on Economic Approaches, Taking into Account the Relationship Between (Interpretation of) Law and Ethics

In a further step, the compatibility of cost-benefit analysis with basic principles of liberal-democratic orders and legal systems must be discussed. If this compatibility is denied, this results in an ethical critique at the same time, as far as one understands these principles as ethical core principles at the same time (which one should do: Sect. 5.1). This kind of criticism is typically not expected in economics. More precisely, if it is, it is expected to be a critique of the self-benefit-centered anthropology and the balancing approach from the legal side. But the two criticisms on balancing and anthropology proved to be less convincing above (Sects. 2.2.1 and 2.2.3). The fact that the legal discourse on cost-benefit analysis or a generally vaguely described economic thinking is often not very productive is also due to this focus. The lamented lack of distributive justice in cost-benefit analysis is also unconvincing insofar as there are only limited constitutional requirements in this regard (Sect. 2.2.6).[1] It is also striking that in the legal literature it is often not very clear what is actually being criticized. The already known problem of mixing cost-benefit analysis and economic instrumentation (Sect. 2.2.8) also exists for a special case here: the legal view of cost-benefit analysis and especially the debate on the economic analysis of law. This is jointly pursued by lawyers and economists and, in addition to economic evaluations, deals with questions of control effects and instruments analyzed in economic terms (either in terms of legal policy or integrated in the interpretation of norms based on the purpose and effects of a norm). Anyway, economic analysis of law is about the integration of economic ways of thinking into the setting and application of law.[2] It is true that economic research can (indeed) contribute substantially to the governance debate on policy instruments. However, this must be separated from the question of whether cost-benefit analysis is compatible with law. The present chapter is focused on this.

It is also unconvincing when some voices in the legal literature oppose economics as an allegedly normative science to law and jurisprudence as an allegedly purely factual subject.[3] What is skewed about this is not only that economics as a discipline is only to a small extent about (normative) cost-benefit analysis and to a larger extent about descriptive questions (such as factual economic developments or the development of governance instruments or legal policy options and legal impact analyses).[4] Above all, it is misunderstood that questions of law (i.e., questions of

[1] As an example see Britz 1997, p. 195 et seq.; see also Grzeszick 2003.

[2] On the debate about economic analysis of law see, e.g., Mathis 2011, p. 3 et seq.; Hammer 2011, p. 211 et seq.; Eidenmüller 2005; Gawel 2001, p. 9 et seq.

[3] As an example Eidenmüller 2005; too vague also Wittmann 2004, p. 34 et seq.; more precisely Winter 2001, p. 97 et seq.

[4] The latter is inaccurately classified as normative by Suchanek 2000, among others, although the statement that a policy instrument is effective in relation to a presupposed goal (possibly balanced against other goals/principles) is undoubtedly a descriptive – instrumentally rational – and not a prescriptive statement.

legal interpretation based on the common legal rules of interpretation of literal meaning, systematics, the purpose of norms and the history of the creation of norms) or the examination of conformity based on the legal norms are themselves normative questions. Facts only come into play as subsumption material in the finding of law.[5] Therefore, law is a kind of special case of ethics. Ethics may lack the concreteness and sanction proof of law, but it may be able to justify the principles behind liberal-democratic legal orders, which law itself cannot do (Sect. 5.1).[6]

It would be equally wrong to reduce the criticism of cost-benefit analysis to the proverbial. The popular dictum "Judex non calculate" (the judge does not calculate), for example, does not bring the necessary precision to the discussion of the principles and application problems of cost-benefit analysis. Moreover, it primarily criticizes quantification (see Sect. 2.2.4 above).

4.2 Claim of Validity of Law and Representative Democracy

It will be shown below that cost-benefit analysis is at odds with core elements of a liberal-democratic legal order. The elements considered are so general that the statement is not limited to a specific national liberal-democratic legal order. The cost-benefit analysis contradicts the constitutional framework of liberal democracies such as Germany, the USA and the EU as a state-like association of states in several respects:

- The liberal-democratic constitutional framework consists of certain rights, whereby freedoms, elementary preconditions of freedom and conditions conducive to freedom are mapped in particular. Furthermore, the framework consists of procedural rules for parliaments, authorities, courts, etc., which are intended to further flesh out these rights and make conflicts between the various rights resolvable. In addition to constitutional balancing rules (including fact-finding rules), which establish a framework for balancing the freedom of one with the freedom of the other, established decision-making procedures play a central role. It is on these, and not on situationally expressed (and in principle constantly changeable) preferences or payments and willingness to pay, as underlies cost-benefit analysis, that political decisions within the constitutional framework are based. Thus, the weight of rights does not depend on how much someone can pay and whether he or she is particularly vocal in demanding his or her rights. Freedom in a liberal democracy is not only the freedom of wealthy consumers with their preferences, but also the freedom of those who are hardly present in

[5] See in detail Ekardt 2019, Ch. 1.6–1.7 and fundamentally Alexy 1991.

[6] On the relationship between law and ethics, see Alexy 1995, p. 127 et seq. and Ekardt 2019, Ch. 1.7: Law always combines normative and instrumental rationality in view of its sanctioning. In my opinion, there is only a limited contrast to Häberle 1974 here, only that what is called constitutional theory there is justified in a different way in the present case. See also Susnjar 2010, p. 34 et seq. and 69 et seq.

the free market due to a lack of purchasing power and who would suffer massively from the consequences of climate change and damaged ecosystems, for example.[7] The irrelevance of the ability to pay for the validity of a right applies even to the freedom of property; in the case of property, too, no other protection is granted legally to a valuable property object than to a less valuable property object. It would not be helpful to raise the objection that it is self-evident that a rich person can make more of his freedom than a poor person, for example, go to meetings more often, express his opinion more easily, and so on. For despite these greater factual chances of realization, the freedom of the rich does not weigh more in legal balancing than the freedom of the poor, and it would be unlawful to give more weight to the rich on the peculiar grounds that they can do more with their freedom anyway. However, what has been said should not be confused with the well-known criticism of utilitarianism that balancing human rights is problematic in itself – for this is precisely not the case (see Sect. 2.2.3).

- The formation of preferences in the context of cost-benefit analysis must inevitably stop at the subjective assessments of individuals, not only in normative judgments but also in the determination of the factual situation, because the approach of cost-benefit analysis rejects any controlling authority over preferences per se. This leads (contrary to the original intentions of an empiricist approach) ultimately also to a relativistic perspective also with respect to the facts, which must treat facts not only as often difficult to determine, but as unknowable in their objectivity. This is not, however, the perspective of liberal democracy, which recognizes a subjective approach to facts only where and to the extent that the facts really cannot be ascertained any better.[8]
- For good reasons, decision-making in liberal democracies is representative-democratic with only isolated plebiscitary (and participatory) additions. Rationality and freedom are – without opening this very complex debate here – best protected in this way, taking into account that human motives are only rational to a limited extent and often problematic in terms of group dynamics (Sect. 2.2.1).[9] Even if one were to assess this last point differently, the representation-oriented status of liberal-democratic constitutions would still be the same. All of this is not invalidated by the fact that representative democracy has historically appealed to some because it keeps the mass of the population at a distance from the immediate political process.[10] Even in isolated liberal-democratic systems, such as Switzerland, where plebiscites are used at all, they are only a supplementary element for a few issues that are the subject of particularly intense public

[7] On the social distributional relevance of climate change, especially for the socially less well off, with further references Ekardt 2019, Ch. 4.6.

[8] Elaborated in detail by Meßerschmidt 2000; Susnjar 2010, p. 300 et seq.; Ekardt 2019, Ch. 3.7.

[9] Steinberg 2013, p. 138 et seq.; Bussemer 2011; Loske 2016, p. 246 et seq.; traced in more detail in Ekardt 2021, § 5 C. II. 3. On the discussion of participatory elements of democracy and their relevance, with further references Ekardt 2010; Fisahn 2004; Ekardt et al. 2012.

[10] On the history of representation, especially since the Vormärz, Scheidler 2015, p. 151 et seq.; Ekardt 2003, p. 15 et seq.; on the legal debate in particular, Steinberg 2013, p. 138 et seq.

debate and are usually not characterized by technical complexity. Here, too, the cost-benefit analysis collides, which wants to link every general (i.e., legal) and individual case-related (i.e., administrative) decision to the changeable factual preferences. The economic discourse about the supposedly greater democratic anchoring of decisions by means of cost-benefit analyses misses this point.[11]

- Particularly important for sustainability issues is the following contrast. Since cost-benefit analyses are based on factual preferences, the concerns of people who do not yet exist cannot appear in them. Likewise, as mentioned, the concerns of those who are not able to pay, and thus of the vast majority of people in developing countries, can hardly be represented. Thus, the exact object of sustainability, namely an intertemporal and globally transboundary broadened perspective on normative questions, is difficult to link with the cost-benefit analysis.[12] In contrast, the law establishes that link to intertemporal and global justice, according to the view expressed here (Sect. 4.3).
- Also for good reasons, confirmed by centuries of historical experience, modern law claims a monopoly of the decision of normative questions for itself, which only knows an opening in favor of other decision-making procedures, such as the free play of forces, where this is permitted by law. A competing normativity to law is thus impossible.[13] The cost-benefit analysis collides with this in that it threatens to establish a competing normativity. This underlines at the same time that the cost-benefit analysis must not ignore the points of criticism shown so far, but must step back behind the law in this respect. Mind you: Even beyond the cost-benefit analysis, any ethics that takes a position on social statements and thus wants to oblige politics to do something would have the problem that the constitutional law of the respective basic political unit (world, EU, nation state, municipality) claims to determine conclusively what politics may and, if necessary, must do, i.e., where its obligations and where its leeway lie. This also results from the insight that ethics cannot achieve the necessary concreteness beyond basic questions to make law dispensable as a mechanism for concrete decisions.[14] Ethics can (this makes it indispensable) provide the basic principles of law with a basis of justification in the first place, check them and thus, in the worst case, prove them to be invalid, but cannot generally simply stand beside law with concrete statements on individual cases (which only become possible through sub-

[11] On the ambivalence also Hanley and Barbier 2009, p. 307; correctly Fehling 2004 with the remark that this allows only the deflated use of the cost-benefit analysis; on this below in Sect. 6.2.

[12] Hanley and Barbier 2009, p. 323 et seq.; Fromm 1997, p. 23 and 258; Hansjürgens and Lienhoop 2015, p. 110 et seq.; on attempts of economic extension in U.S. practice, which then, however, leave the approach of cost-benefit analysis and its basis of justification by arbitrary assumptions about the preferences of, e.g., future people, also Gayer and Viscusi 2014.

[13] See pars pro toto Vöneky 2010, p. 107 et seq.; Ekardt 2019, Ch. 1.7 and 3.1; Ekardt 2021, §§ 1 D. III. 3., 3 A.

[14] This is explained in detail in Ekardt 2019, Ch. 3. It should be noted that questions of legal interpretation are also disputed. Nevertheless, (a) the practical relevance, (b) the restriction of the possible spectrum of opinions and (c) a decision-making mechanism for disputes in the form of democracy and separation of powers are relatively clearly given here.

jective determinations on the part of the legislator). Otherwise, one would end up exactly where Hobbes had ended up in the seventeenth century: Hobbes (see already Sect. 2.1.1) feared sheer civil war due to competing normative views; if this were allowed to happen, law would ultimately lose its meaning. Hobbes' answer to this problem was ultimately to set aside ethics in favor of law. While this is only convincing from a skeptical ethical position, which is explicitly not advocated here (Sect. 5.1), competing normativities are just as unconvincing, especially since they would undermine precisely central ethical principles in the form of legal certainty and the elementary precondition of freedom: peace.[15] Compared to ethics, cost-benefit analysis cannot even offer a justificatory basis for law because it will prove to be an unconvincing ethics approach later in Sect. 5.1.

Thus, a liberal-democratic right has concepts of rights, fact-finding, democracy, and decision-making monopolies that contradict cost-benefit analysis and, in doing so, do not tolerate any alternative concept alongside them.

4.3 Legal Balancing Rules as a Means of Limiting the Scope of Political Decision-Making – And the Extent to Which They Contain Cost-Benefit-Analytical Yardsticks

Building on this, it must be briefly explained here what a liberal-democratic legal decision-making scheme for conflicts looks like. This is also necessary because it makes visible which aspects of cost-benefit thinking are already contained in such a legal system, ergo do not contradict it. The question of whether the following also reflects the ethical standards for social conflicts is not discussed further here. According to the opinion represented here, this is the case, because the rules would fit exactly to the ethics approach to be briefly presented in Sect. 5.1. However, nothing would change for the criticism of cost-benefit analysis if this were not true.

The basic problem from the point of view of a liberal-democratic legal system is the resolution of conflicts between conflicting interests. Balancing is thus inevitable, as already shown (Sect. 2.2.3), especially in environmental protection, but also otherwise in principle. By allowing the industrial society, by approving industrial plants, by permitting car traffic, etc., there is statistically more or less inevitably collateral damage, i.e., impairment of the right to life and health, due to the released air pollutants, etc. This happens in balancing with the human rights to freedom of the consumers and the companies. As mentioned, there is no general formula "Neminem laedere" – almost everything in life is disadvantageous for someone (see

[15] On these principles, which are not to be discussed in detail here, Ekardt 2021, §§ 4 C., 5 C. II. 2.; Ekardt 2019. The classic question of the natural law-positivism debate over the centuries, whether law can suddenly appear so incorrect that it becomes directly invalid and thus must give way to ethics in direct application, need (arguably) not be decided for the present problem context; more on this question with further references Ekardt 2019, Ch. 1.7.

already Sect. 2.2.3).[16] It is not for nothing that human rights are incontestably restrictable.

For those involved in the legal discourse, it is important to clarify that balancing in this case refers comprehensively to the question of a fair balancing of conflicting interests, regardless of whether it is a matter of legislation (which represents a balancing by the legislator) or applying the law (with official or judicial balancing), and regardless of whether, in the case of applying the law, the balancing of interests is a matter of interpreting the legal terms of a norm or of a discretionary review, proportionality review or planning decision.[17] Ultimately, any legislation and application of the law must simultaneously attempt to do justice to the conflicting rights to freedoms and other normatively recognizable concerns of the people behind them, and thus represents precisely a balancing decision. All of this can therefore be traced back to the constitutional requirement that any legislator (be it regional, national or transnational) must balance conflicting interests in a fair and just manner.[18]

So how exactly is human rights-based environmental protection, specifically the right to the elementary preconditions of freedom – life, health and minimum subsistence (including their intertemporal and global cross-border dimension: Sect. 5.1) – to be reconciled with the fundamental rights of companies and consumers at the various legal levels? And what democratic leeway is left for politics – how much

[16] Gawel 2001, p. 9 et seq.; Lübbe 2000, p. 73 et seq.; Winter 2001, p. 97 et seq.; Ekardt 2019, Ch. 3.5.

[17] Legal interpretations of norms are carried out according to the four so-called canons of interpretation, namely the (grammatical) interpretation according to the wording of the norm, the (systematic) interpretation according to the location and context of the norm in relation to other norms, and – according to a controversial view that may require qualification – also according to the discernible purpose of the norm (teleological) and according to its history of origin (historical); in detail, with further references Ekardt 2019, Ch. 1.7 and Ekardt 2021, § 1 D. III. 4. The question of whether this is based on a more semantic or a more pragmatic or an even different understanding of language is not addressed in the legal discourse (on this distinction Meggle and Kompa 2011, p. 203 et seq. and on a detailed mediating approach Meggle 2010, p. 55 et seq.). In a way, the dispute on realism versus pragmatics in semantics seems like an echo (albeit of course shifted) of the medieval dispute on universals (see above Sect. 2.1.1). The question of realist or pragmatic semantics was already being considered in the fourteenth century; see Beckmann 2010, p. 28.

[18] The framework of legislative balancing is usually called the proportionality test. For the administration, on the other hand, where the legislator has already carried out this balancing to a considerable extent, the balancing – often without being legally designated as such – is limited to the interpretation (never free of leeway) of the facts of the norms that the legislator has created as an expression of his balancing, as well as to the filling of explicitly so-called discretionary or planning balancing leeway. This entire basic structure applies cum grano salis irrespective of whether we are talking about national, European or international law. On the entire problem in more detail with further references Calliess 2002, p. 373 et seq.; Susnjar 2010, p. 199 et seq.; Ekardt 2019, Ch. 3.5–3.6. The (intertemporal and transboundary) environment-related human rights protection considered in more detail elsewhere therefore does not apply without balancing, just as this can also be said of human rights guarantees in general. In addition to the fundamental balancing necessities already mentioned, this already results from the point that environmental and, more specifically, climate protection itself interferes with citizens' guarantees of economic freedom and is thus limited under human rights law.

climate protection, for example, is ultimately required in concrete terms? The formal requirements are important, e.g., in the form of the allocation of responsibilities, but so are the substantive rules for balancing the various options, which are commonly known by the term proportionality, which tends to obscure the situation. They indicate the margins of discretion and lead to the answer to the question of whether, for example, more can be demanded in terms of environmental protection under human rights law, so that, on the one hand, human rights protection does not run dry, but, on the other hand, courts cannot intervene in democratic decisions at will and make policy themselves, but can only demand as much human rights protection as the limits set by the formal rules and balancing rules demand.

Which formal limits and substantive balancing rules apply in detail – and precisely leave a considerable scope for decision-making – can already be derived in essence from human rights themselves, for "defense" against public authorities and "protection" by those authorities against the fellow citizens alike (although legal practice refers to this to varying degrees).[19] This derivability from human rights becomes apparent first for the basic balancing rule, which in the usual conceptualization of balancing as a proportionality test is usually addressed under the heading of "legitimate purpose": that the balancing material must be complete and may not contain any illegitimate concerns (such as, for example, protection against oneself). If the assumption, justified in more detail elsewhere, is correct that self-determination or the newly interpreted freedom – and everything that follows from it in terms of elementary preconditions of freedom as well as further conditions conducive to human-rights freedom (e.g., the protection of biodiversity) – is the only possible object of regulation of state action, it is also relatively easy to state as a rule of balancing which is the (sole) permissible material of constitutional balancing. If one considers this not to be the case, one would generally have to speak of public and private interests being possible objects of balancing.[20] In essence, human rights protection, e.g., against climate change, involves a collision of the right to the elementary preconditions of freedom with the economic rights to freedom, as they are regulated, for example, in Germany in Articles 2, 12 and 14 of the German constitution or inter- and supranationally in the European Charter of Fundamental Rights and the International Covenant on Civil and Political Rights. It also follows from the human rights guarantees of freedom themselves that orders of competence and procedural rules, such as those governing the legislative process, must be respected. In general, the rules of deliberation (which can also be derived from the guarantees of freedom and are usually explicitly written down in constitutions) also contain procedural rules that contain a responsibility for such decisions and, ergo, for filling the

[19] On this and the following, Calliess 2002, p. 373 et seq.; Susnjar 2010, p. 199 et seq.; the following aspects of freedom and balancing rules are dealt with in detail in Ekardt 2019, Ch. 3 (which is why the following presentation is limited to a very brief reproduction).

[20] On this and the following with further references Ekardt 2019, Ch. 3.4 (there also on the authoritarian, in part even totalitarian legal history of the concept of common good or public interest); more openly Häberle 1974.

remaining margins, ergo, that establish the democratic institutional order based on the separation of powers.[21]

Furthermore, the balancing rules well known as the proportionality test, i.e., the appropriateness and necessity of a restriction of liberty in favor of the interests of other parties, follow directly from human rights, even if they are little discussed nationally and internationally for rights protecting freedom by the state against the fellow citizens: One person's freedom may not be taken away if no other person's freedom benefits from it. Under the rubric of the appropriateness of a state action protecting environmental human rights (and abridging the environmental human rights guarantees), it can be asked, for example, whether national climate protection measures can have any relevant effects at all, given that climate protection is a global problem. On closer inspection, however, the objection to climate protection based on human rights is not very valid. First, one legal consequence of environmental human rights protection may be that there is an obligation to cooperate globally. And secondly, national measures that are potentially imitated by others are not devalued by the fact that the rest of the world does not immediately follow suit.

Reasonableness, as the final step in the traditional proportionality test, can also be seen as the umbrella for a number of other balancing rules that also follow from the conflicting human rights guarantees. One of these rules is that one concern must not be evidently set aside too unilaterally in favor of other concerns; again, this follows from the idea that freedom as a whole should be maximized (but even with relatively extensive climate protection, noise protection, etc., deaths and damage to health will still occur). Another component of appropriateness is, of course, that human rights guarantees must be included in the balancing process according to the degree to which they are actually affected.

Another balancing rule that can also be examined under the heading of appropriateness is the polluter pays principle or the junction of freedom and responsibility for the consequences of one's actions, which again can be derived from human-rights-based freedom itself: For freedom must include a responsibility for foreseeable consequences. The negative effects of an action that is otherwise positive for the actor (e.g., cheap free transportation today) must therefore, at least in principle, be allocated at the actor, e.g., by way of charging for ecological damage. So far, this is missing in large parts of environmental protection and climate protection. This polluter-pays relationship is not cancelled out by the fact that all other people or states worldwide also cause greenhouse gas emissions, because violations of human rights are not rendered irrelevant by the fact that others also commit them. Another obvious balancing rule arises from the need to finance the protection of the preconditions of freedom and from the idea contained in the junction of freedom and responsibility for consequences that the community should, as far as possible, compensate for problems that someone has through no fault of his own. and from the idea of performance contained in the junction of freedom and responsibility for consequences. This is explicitly standardized, for example, in Article 13 of the

---

[21] See in detail with further references Ekardt 2019, Ch. 3.5–3.6; Ekardt 2021, § 5 C. II. 3.; Susnjar 2010, p. 199 et seq.; Calliess 2002.

French Declaration of the Rights of Man and of the Citizen, but other constitutional systems, such as Germany's, also take it as an implication of the rights of freedom.

The polluter pays principle and the ability to pay can become relevant in determining the correct way to deal with the climate problem, but they can point in different directions. This, together with the relative modesty of the aforementioned balancing rules, leaves the political authorities with a great deal of decision-making leeway in dealing with climate change. This changes fundamentally, however, if one recognizes a further balancing rule that has not yet played a big role in practice: As an outflow of the human-rights-based freedom (and its preconditions), it is obvious that the political scope for decision-making ends where a political action or omission substantially endangers the liberal-democratic system as a whole. And this is precisely the case if drastic steps are not taken promptly to protect the climate. Thus, drastic climate protection – subject to the handling of uncertain factual situations, which will not be discussed in more detail here – appears to be fundamentally imperative.

In view of the limited capacities of representative democracy in times of complex and numerous political decisions, environment-related human rights also imply procedural guarantees, in particular a right to appropriate participation in administrative and indirectly, for example, through the use of the Internet, also in legislative decisions. Furthermore, it is important to have rules for ascertaining the facts on which balancing is based, including how to deal with uncertain facts, which is typical of sustainability issues.

It is quite obvious that rules such as the appropriateness and necessity rule have points of contact with the cost-benefit analysis. The same applies to the fact that balancing is clearly carried out at all. In the liberal-democratic legal framework, however, it remains a qualitative framework, and neither the balancing nor the procedure follow the path of plebiscitary permanent quantification.[22] One could also put it this way: In contrast to the legal or ethical balancing theory proposed here, cost-benefit analysis does not aim to sound out the limits of normative rationality and, on the basis of this, to provide political decision-makers with a framework for their balancing leeway. Rather, cost-benefit analysis aims from the outset at exactly one concrete recommendation as to how much climate protection, water protection, resource conservation, etc. should be pursued politically. This claim, however, already proved to be unrealizable up to this point - already disregarding questions about the justification-theoretical basis of the cost-benefit analysis.

The contrast thus diagnosed between cost-benefit analysis and liberal democratic legal order does not contradict the fact that cost-benefit analysis is sometimes practiced in liberal democracies such as the USA or Great Britain. Indeed, this practice exists in the Anglo-Saxon area and sometimes also by the EU when new legislative

[22] Aptly also Winter 2001, p. 97 et seq.; Meßerschmidt 2001; Martini 2008, p. 208 et seq.; Eidenmüller 2005, p. 468 et seq.; auch Britz 1997, p. 185 et seq.; Künzler 2012; partly ignored by Führ 2003, p. 257 et seq.

acts are planned.[23] But apart from the fact that this is done only selectively there as well (for example, in the case of waterway improvements or road construction measures[24]), the use there can be interpreted similarly to a non-binding referendum that does not develop any precedent-setting force for the ultimate decision. In part, cost-benefit analysis also provides real facts, when it is not a matter of hypothetically determined preferences (for example, via willingness-to-pay analyses or supposedly revealed preferences derived from other actions). This can be, for example, about impending losses of goods with real market prices. Such facts are also of great importance according to constitutional standards in balancing processes and must therefore considered because without them it is not possible to determine how strongly certain rights are really affected. At this point, it becomes clear why it was mentioned at the beginning of the book that a critique of cost-benefit analysis leaves open the possibility of continuing to pursue it in a deflated, methodologically clearly limited manner. How this can be meant exactly, will be considered at the end of the volume.

Repetition Questions

1. What are the common misconceptions about the relationship between law and economics? (Sect. 4.1)
2. How does cost-benefit analysis relate to the claim of validity of law? (Sect. 4.2)
3. How does cost-benefit analysis relate to human rights and representative (not plebiscitary) democracy? (Sect. 4.2)
4. How does a legal balancing approach differ from cost-benefit analysis, and what are some important balancing rules? (Sect. 4.3)
5. How can a duty to protect the climate be derived from human rights? (Sect. 4.3)

References[25]

Alexy R (1991) Theorie der juristischen Argumentation, 2nd edn. Frankfurt a.M

Alexy R (1995) Recht, Vernunft, Diskurs. Frankfurt a.M

Beckmann J (2010) Wilhelm von Ockham, 2nd edn. Beck, C H, Munich

Britz G (1997) Umweltrecht im Spannungsverhältnis von ökonomischer Effizienz und Verfassungsrecht, Die Verwaltung:185 et seq.

Bussemer T (2011) Die erregte Republik. Wutbürger und die Macht der Medien. Klett-Cotta, Stuttgart

23 COM, SEC (2005) 791; Hofmann 2007, p. 50 et seq.; Winter 2001, p. 97 et seq.; Hanley and Barbier 2009.

24 See on the respective debates Lange and Krull 2014; Fehling 2004; Gawel 2014; Ekardt 1997.

25 In accordance with legal practice, parliamentary, governmental and EU Commission documents as well as laws and judgments are not listed in the bibliography, as they can be found unequivocally based on the reference given in the continuous text or via the general search engines. The last access date for all Internet sources is 31.10.2021.

Calliess C (2002) Rechtsstaat und Umweltstaat. Mohr Siebeck, Tübingen
Eidenmüller H (2005) Effizienz als Rechtsprinzip. Möglichkeiten und Grenzen der ökonomischen Analyse des Rechts, 3rd edn. Mohr Siebeck, Tübingen
Ekardt F (1997) Zur Verfassungsmäßigkeit der sogenannten Privatfinanzierung von Verkehrswegen. Verwaltungsblätter für Baden-Württemberg:281 et seq
Ekardt F (2003) Liberalismus, Besitzindividualismus und Handlungstheorie. Leipziger Universitätsverlag, Leipzig
Ekardt F (2010) Information, Partizipation, Rechtsschutz. Prozeduralisierung von Gerechtigkeit und Steuerung in der Europäischen Union, 2nd edn. Münster
Ekardt F (2019) Sustainability. Transformation, governance, ethics, law. Springer, Dordrecht
Ekardt F (2021) Theorie der Nachhaltigkeit. Ethische, rechtliche, politische und transformative Zugänge – am Beispiel von Klimawandel, Ressourcenknappheit und Welthandel, 4th edn. (= 3rd ed. of the new edition). Baden-Baden
Ekardt F, Heitmann C, Susnjar D (2012) Sicherung sozial-ökologischer Standards durch Partizipation. Düsseldorf
Fehling M (2004) Kosten-Nutzen-Analysen als Maßstab für Verwaltungsentscheidungen. Verwaltungsarchiv:443 et seq
Fisahn A (2004) Demokratie und Öffentlichkeitsbeteiligung. Mohr Siebeck, Tübingen
Fromm O (1997) Möglichkeiten und Grenzen einer ökonomischen Bewertung des Ökosystems Boden. P. Lang, Frankfurt a.M
Führ M (2003) Eigen-Verantwortung im Rechtsstaat. Berlin
Gawel E (2001) Ökonomische Effizienzforderungen und ihre juristische Rezeption. Ein problemstrukturierender Überblick. In: Gawel E (ed) Effizienz im Umweltrecht. Grundsatzfragen einer wirtschaftlichen Umweltnutzung aus rechts-, wirtschafts- und sozialwissenschaftlicher Sicht. Nomos, Baden-Baden, p 9 et seq
Gawel E (2014) Konzept zur „Berücksichtigung“ von Umwelt- und Ressourcenkosten nach Art. 9 WRRL, Verwaltungsarchiv:99 et seq
Gayer T, Viscusi K (2014) Determining the proper scope of climate change benefits. Manuscript
Grzeszick B (2003) Lässt sich eine Verfassung kalkulieren? Juristenzeitung 58:647 et seq
Häberle P (1974) Verfassungstheorie ohne Naturrecht. Archiv des öffentlichen Rechts 99:437 et seq
Hammer B (2011) Valuing the invaluable? Valuation of human life in cost-efficient assessments of regulatory interventions. In: Mathis K (ed) Efficiency, sustainability, and justice to future generations. Springer, Berlin, p 211 et seq
Hanley N, Barbier E (2009) Pricing nature. Cost-benefit analysis and environmental policy. Elgar, Cheltenham
Hansjürgens B, Lienhoop N (2015) Was uns die Natur wert ist. Potenziale ökonomischer Bewertung. Metropolis-Verlag, Marburg
Hofmann E (2007) Abwägung im Recht. Chancen und Grenzen numerischer Verfahren im öffentlichen Recht. Mohr Siebeck, Tübingen
Künzler A (2012) The jurisprudence of welfare maximization. Embracing complexity and uncertainty in the environmental, health and safety context. Archiv für Rechts- und Sozialphilosophie 98:1 et seq
Lange C, Krull D (2014) Infrastrukturelle Wasserdienstleistungen. Zur Funktion kostendeckender Wasserpreise nach Art. 9 der EG-Wasserrahmenrichtlinie und ihre mögliche Anwendung auf den Aufstau von Gewässern als infrastrukturelle Wasserdienstleistung für Binnenschifffahrt und Wasserkraft. Metropolis-Verl, Marburg
Loske R (2016) Politik der Zukunftsfähigkeit. Konturen einer Nachhaltigkeitswende. Fischer Taschenbuch, Frankfurt a.M
Lübbe W (2000) Neminem laedere? Archiv für Rechts- und Sozialphilosophie. Beiheft 74:73 et seq
Martini M (2008) Der Markt als Instrument hoheitlicher Verteilungslenkung. Mohr Siebeck, Tübingen
Mathis K (2011) Discounting the future? Cost-benefit analysis and sustainability. In: Mathis K (ed) Efficiency, sustainability, and justice to future generations. Springer, Berlin, p 165 et seq
Meggle G (2010) Handlungstheoretische Semantik. Berlin et al, De Gruyter

Meggle G, Kompa N (2011) Pragmatics in modern philosophy of language. In: Bublitz W, Norrick N (eds) Foundations of pragmatics. DeGruyter Mouton, Berlin et al, p 203 et seq

Meßerschmidt K (2000) Gesetzgebungsermessen. Berlin

Meßerschmidt K (2001) Ökonomische Effizienz und juristische Verhältnismäßigkeit – Gemeinsames und Trennendes. In: Gawel E (ed) Effizienz im Umweltrecht. Grundsatzfragen einer wirtschaftlichen Umweltnutzung aus rechts-, wirtschafts- und sozialwissenschaftlicher Sicht. Nomos, Baden-Baden, p 215 et seq

Scheidler F (2015) Das Ende der Megamaschine. Geschichte einer scheiternden Zivilisation. Promedia, Wien

Steinberg R (2013) Die Repräsentation des Volkes. Menschenbild und demokratisches Regierungssystem. Nomos, Baden-Baden

Suchanek A (2000) Normative Umweltökonomik. Tübingen, Mohr Siebeck

Susnjar D (2010) Proportionality, fundamental rights, and balance of powers. Martinus Nijhoff Publishers, Leiden

Vöneky S (2010) Recht, Moral und Ethik. Grundlagen und Grenzen demokratischer Legitimation für Ethikgremien. Mohr Siebeck, Tübingen

Winter G (2001) Über Nutzen und Kosten der Effizienzregel im öffentlichen Recht. In: Gawel E (ed) Effizienz im Umweltrecht. Grundsatzfragen einer wirtschaftlichen Umweltnutzung aus rechts-, wirtschafts- und sozialwissenschaftlicher Sicht. Nomos-Verl.-Ges, Baden-Baden, p 97 et seq

Wittmann BK (2004) Effizienzsteigerung bei Gerichten im Spannungsfeld zwischen richterlicher Unabhängigkeit und Anwendung betriebswirtschaftlicher Methoden. Munich

Cost-Benefit Analysis Without Convincing Theoretical Basis

5

Abstract

The empiricist theoretical basis of (covertly practical-philosophical, normative) cost-benefit analysis is untenable. This theoretical basis consists of the empiricism-oriented idea that normative questions have to be transformed into questions of factual preferences. In this context, there are logically false conclusions and performative contradictions, among other things because the idea that normative questions per se are not objectively decidable leads to self-contradictions. Furthermore, cost-benefit analysis does not hold up to any application to itself and, upon closer examination, can hardly make plausible how it actually fits its own empiricist foundations. Cost-benefit analysis also fails to find a convincing approach to intertemporal justice. Moreover, the economist attempt to deflate ethics into a question of factual preferences for altruism is unsuccessful. Similar problems also arise for so-called economic ethics, which, apart from monetization, just runs largely parallel to cost-benefit analysis.

5.1 Theoretical Frictions of Cost-Benefit Analysis as Empiricist Ethics – And (Heterodox) Discourse-Ethical Alternatives

After application problems and compatibility problems with liberal-democratic (legal) basic principles, the justification-theoretical basis of cost-benefit analysis is now put to the philosophical test. This refers to the empiricist understanding of rationality, which is limited to quantifying balancing and instrumental (as well as

F. Ekardt, *Economic Evaluation, Cost-Benefit Analysis, Economic Ethics*, Environmental Humanities: Transformation, Governance, Ethics, Law 2, https://doi.org/10.1007/978-3-030-99284-2_5

theoretical) questions, but leaving aside normative questions.[1] The suggestion that quantifying cost-benefit analysis contradicts concepts of liberal-democratic constitutions could possibly be answered by economists by doubting the rightness or rather (in their words) the efficiency of liberal-democratic law. But perhaps the liberal-democratic legal system has a convincing ethical basis. This kind of critical demand is traditionally not expected in the debate of economic science, since such points of view are largely impresent there. Rather, it is apparently simply assumed that an empiricist basis with consumer sovereignty, preference orientation, etc. is simply an unquestionable scientific world view. Whether this is correct or not is now to be analyzed. This shows at the same time that the liberal-democratic basic principles of the legal order do indeed have an ethical foundation – and that empiricist non-cognitivism is exposed to serious objections in normative questions.[2] The existence of an objectively founded ethics, moreover, avoids, as mentioned, that problems of comparability of ethics and legal order arise. At the same time a meaningful complementary relationship emerges, in that ethics does not need to compete with the proof of sanction (which it cannot possibly do) or concreteness of law (which, in view of the immense scope of the legal order, it can hardly do either), and yet clearly assumes an outstandingly central role.

It should be recalled: When cost-benefit analysts deny the possibility of objective statements for normative questions (in contrast to questions of fact) in agreement with the majority of other experts such as natural scientists, sociologists and political scientists, they follow a skeptical or positivistic ethical position with their empiricist conception of rationality. The fact that this (contrary to the self-perception of most economists) represents a normative or ethical position at all has already been considered (Sect. 2.2.2). Cost-benefit analysts consider only the rational balancing between these preferences to be possible. However, as seen, this balancing is not experienced as a normative act, but rather as a quantifying accounting of diverging preferences. This idea of optimality or efficiency can also be found among less conventional economists.

Central objections speak against the preference theory and its narrow, empiricist concept of rationality (also in the case of climate protection). Here, the aspect is not taken up again that such arguments could also be presented in a similar form against contextualist – i.e., based on cultural traditions or moral intuitions – or postmodern ethical positions.[3] Specifically, contextualist positions clearly differ from

[1] Here, the old controversy may remain open, whether such a reflection within philosophy is to be classified to theoretical philosophy (in the form of epistemology) or to practical philosophy (in the form of ethics respectively theory of justice; even further designations are conceivable). Nothing depends on this systematizing classification in terms of content.

[2] More detailed on the following own justificatory philosophical position Ekardt 2021, § 3 F.-G.; Ekardt 2019, Ch. 3.1; partly similar (and in some points inspired by) Alexy 1995, p. 127 et seq.; Illies 2003, p. 129 et seq.; Kuhlmann 1985; Habermas 1983, p. 56 et seq.; implicitly also Ott and Döring 2004, p. 91 et seq. and Rogall 2012, Ch. 3.1.

[3] See with further references Ekardt 2021, §§ 3 C., 3 D.; Susnjar 2010, p. 219 et seq. In any case, the universalistic perspective presented here suggests that economists cannot escape the following argumentation in a Parsonsian (or ultimately constructivist: see above Sect. 1.4) manner by simply claiming that every scientific discipline is based on some initial dogmas that constitute the discipline and are in any case unquestionable for outsiders.

cost-benefit analysis in that they consider only altruistic preferences, so to speak.[4] On the other hand, they are also based on factual assessments of people (which generates the following questions to cost-benefit analysis practically throughout as well, supplemented by the point that the shared moral intuitions are mostly only asserted, whereas economists at least put considerable effort into the empirical assessment of their preferences, even if they fail with it: Sect. 3.1).[5] In particular, it deals with the following aspects:

- Our purely factual will is correct per se according to preference theory. Preferences take the place of what in ethics are traditionally rights, duties or principles. Their critical examination and justification is thus per se no longer possible. Economists could perhaps reply that this is unusual or regrettable, but unavoidable. But we are not only faced with a practical problem, but also with a logical one. For there is an is-ought fallacy: From the fact that I wish something, it cannot simply follow that this is automatically normatively correct.[6] With an axiomatic statement (preferences are per se correct) this cannot be caught up, because this statement would lose its basis with the simple question: why. And

[4] The line pursued by Lübbe 2015, p. 30 et seq. to insinuate different categories in this respect is not convincing in my opinion.

[5] Lübbe 2015; Lübbe 2016, p. 325 et seq.; Lübbe 1998 and Betzler and Schroth 2014, p. 279 et seq. also criticize the cost-benefit analysis, partly with arguments similar to the present one, but partly also with arguments that are, in my opinion, not convincing (see Ch. 2.2.3 and 2.2.4). And some of them omit essential arguments that cannot be derived from their contextualist or moral intuitionist approach. Beyond the criticism formulated in Ekardt 2019, Ch. 3.1 one might ask how intuitionists intend to deal with the fact that their approach potentially characterizes our Stone Age attitude toward things (Wilson 2012; Ekardt 2019, Ch. 2.6) as per se correct, especially our altruism, which is largely limited to our own group and quite limited there as well – in a world whose conflicts go far beyond the geographical structures of the Stone Age. Exactly this "stone-age-ness" of human intuitions turned out to be a major reason why climate change is not adequately addressed. The intuitions in this regard have remained with us until today, despite all complex cultural overlays. Furthermore, one could ask whether moral intuitions – although they have not yet been carefully surveyed – mean our expressed attitudes or rather our attitudes that become visible in actions. This makes a big difference, since, for example, our talk is far more ecological than our actions (on this with further references Ekardt 2019, Ch. 2.2). The cost-benefit-analysts at least take up the associated methodological problems to some extent. The preference-based, subjective, naturalistic thinking is also criticized by von Kutschera 1982, p. 126 et seq. and passim. However, the intuitionism there based on so-called value experiences appears as a form of the just criticized contextualism. Among other things, it repeatedly argues with everyday intuitions, e.g., that certain norms are valid without balancing, but do not stand up to closer scrutiny; for this example, see above Sect. 2.2.3. Moreover, in a value-experience approach it remains indissolubly unclear what the rightness of the respective perceived (factual) value experience is to be determined by. Purely factually, everyone has value experiences, but often very different ones. The author, for example, has different ones than Angela Merkel, Heinrich Himmler and Mother Theresa; and these three also have (obviously) quite different value experiences in themselves.

[6] This is also noted by Hampicke 2011. Another question is whether positions that are not empirically realizable under any circumstances, for instance for reasons of natural law, are therefore perhaps also normatively wrong ("ultra posse nemo obligatur"); this has been discussed elsewhere; see Ekardt 2019, Ch. 3.1.

the idea that the preference approach is without alternative because normative questions cannot be decided with reasons and therefore recourse must be taken to factual preferences, does not justify the approach either. For first, it will be shown in the following that there is a constructive rationality-theoretical alternative. And second, even if there were no constructive alternative, because normative questions really could not be decided with reasons, it would still be conceivable to then simply take a skeptical path and give up formulating (reasoned) normative standards altogether.

- If one argues for consumer sovereignty, the further question arises: Should it be about average utility, about utility sums or about a genuine consensus?[7] Or may the majority simply be allowed to make arbitrary decisions based on their factual preferences, without being hindered by any frameworks (such as those provided by liberal democracy in the form of guarantees of freedom)? And if this is the case: Is 50.1% of a society allowed to make arbitrary decisions, or 73.4%, or rather 84.5%? It is not apparent how one would answer these questions other than arbitrarily. In this respect, the preference approach is not only (see above) inconclusive, but also underdetermined in central respects and thus ultimately meaningless.
- A further objection can be drawn from the considerations of liberal democracy in Chap. 4. If factual preferences are important, why does the cost-benefit analysis not follow the factual preferences of the population in Western states for a constitutional democracy with separation of powers, which precisely does not provide for plebiscite-like individual case decisions based on individual case-oriented preference determination? No real answer seems possible to this question either.
- The cost-benefit analysis is officially based on an economic Kaldor-Hicks test and yet at the same time violates its fundamentals. The object of the Kaldor-Hicks test is actually to identify situations to which even the losers will voluntarily agree for the sake of compensation. Any serious Kaldor-Hicks test, however, is bound to fail in practice insofar as it deals with situations involving many parties (as is simply always the case in politics). This is because the Kaldor-Hicks test gives each everyone a veto position. Even a single hardcore climate activist who, for example, strictly rejects any further emissions can thus make any solution based on the Kaldor-Hicks criterion impossible – and so can a climate sceptic.[8]
- Furthermore, as already noted when discussing the problems with the law and in the questions of application, dealing with the concerns of future generations

[7] See Harsanyi 1978; Smart 1978, p. 111; Singer 1979; Crane and Matten 2010, p. 98 et seq.

[8] Hanley and Barbier 2009, p. 307 and 319 et seq.; Pissarskoi 2014, p. 101 et seq. and 127 et seq.; more mathematically elaborated are further ambiguities in Breyer and Kolmar 2014, p. 65 et seq.; Eidenmüller 2005, p. 155 et seq.; also implied in Pearce et al. 2006, p. 32 et seq. A different argument with the same basic intention is formulated by Spangenberg 2016–2017, when he argues that cost-benefit analysis can only work if the ideal state of equilibrium against which everything is to be measured is known. In my opinion, such an "internal" critique is ultimately not convincing, because neoclassical economics postulates precisely that factual preferences would automatically lead to an equilibrium of supply and demand.

poses major problems for cost-benefit analysis, or more precisely, unsolvable problems.[9] According to the basic ideas of cost-benefit analysis, future people cannot have preferences and are therefore irrelevant. If they are nevertheless considered, the theoretical basis of cost-benefit analysis is destroyed. However, if they are not considered, this is not only legally and ethically unacceptable (see below). It also destroys the foundations of cost-benefit analysis, which, according to the Kaldor-Hicks criterion, requires compensation for the injured. This, however, cannot be reasonably achieved intergenerationally or can only be achieved in such a way that the consistency of the cost-benefit analysis suffers, e.g., by an externally imposed limitation of free preferences.[10] This is not a marginal phenomenon. Rather, the cost-benefit analysis in environmental protection and especially in climate protection always involves strong intertemporal relations in every decision (whether of a fundamental or project-related nature). Consequently, nothing can be said about the extent of the necessary climate protection, nature conservation, soil protection, etc., if this intertemporal axis is ignored. The aspect that the rights to the elementary preconditions of freedom (life, health and subsistence, which can be weighed up like all rights) also have an intertemporal effect has been discussed in detail elsewhere.[11] One argument was related to the character of liberties as protection where they are threatened: namely, increasingly also intertemporally (and globally across borders). Another argument arises from the basic ethical approach developed in the second part of this section.

- Furthermore, the application of the theory to itself is likely to fail in the cost-benefit analysis. Assuming that its proponents adhere to the theory themselves, one would have to assume that they ultimately advocate the theory based on their own factual preferences, whereby these preferences would have to be assumed to be primarily selfish. This raises further doubts about cost-benefit analysis, at least prima facie. In other words, if the normative and empirical notion of cost-benefit analysis is correct, that people consistently act in a conscious and calculated manner, essentially self-interestedly, and that this is also normatively to be welcomed, then economists, for example, will also prefer to make such recommendations as will secure them further research contracts. To put it ironically: The methodologically hardly successful "figure orientation," which is popular with politicians and the media, may in this case serve the factual preferences of cost-benefit analysts, but nothing more.
- Moreover, the preference theory contains a self-contradiction. For whoever says that there are no general normative propositions, and therefore preferences must be taken as a general basis, is himself making a general statement about norms.

[9] Hanley and Barbier 2009, p. 323 et seq.; Fromm 1997, p. 258 and passim; Gayer and Viscusi 2014.

[10] Fromm 1997, p. 248 et seq.; Hanley and Barbier 2009, p. 323 et seq.

[11] See with further references Ekardt 2019, Ch. 3.3; Unnerstall 1999.

The statement "everything is only subjective in norms" thus refutes itself.[12] Objective morality is logically just not deniable in its possibility; its denial contradicts itself.[13] Thus, we humans logically necessarily presuppose that normative questions can be decided at all with reasons and ergo objectively and not only subjectively preference-driven; otherwise, we contradict ourselves. We even presuppose this implicitly every day (even if we never explicitly engage in a discourse of reasons about that question) when we pose normative theses in our conversations and give reasons for them, i.e., mark them with the claim of objective visibility, instead of presenting them only as subjective. And it should be almost impossible never to formulate sentences with "because, there, therefore" to normative questions throughout one's life. Thus, there is no escape from the fundamental possibility (!) of objectivity in normative questions. But we also logically presuppose the possibility of objective statements when we say: "I am a sceptic and say that objectively there are only subjective statements about ethics." This statement can only be valid if there is objectivity after all. Thus, the critique of objectivity logically cancels itself out. (Since the cost-benefit analysts themselves present a normative theory, albeit unintentionally, it is even irrelevant for the pure refutation of the cost-benefit analysis whether a strict skepticism would be tenable.) The fallibility of human knowledge is not questioned by this. Also that even objective (one could also say: intersubjective) cognition – with facts and with norms – always cannot exceed the human limits of cognition, is not at all in question. It is beyond the scope of this book to show that what has

[12] Elenctic/negative/transcendental pragmatic arguments of the following kind are used particularly by Alexy 1995, p. 127 et seq.; Illies 2003, p. 129 et seq.; Kuhlmann 1985; Ott and Döring 2004, p. 91 et seq. and passim. The structure of the negative (precisely not deductive) argument, with which one just avoids an infinite regress or an arbitrarily set basic axiom, however, is already known since the times of Plato, Augustine, and Thomas Aquinas (as a logical figure, not concrete to the subject matter treated in the text). For some misunderstandings that often occur in the philosophy/ economics discourse, see the dispute between Dilger 2006 and Ekardt 2006.

[13] This point of criticism could be formulated as a more general criticism of philosophical empiricism by also looking at the empiricist ideas about knowledge of facts (i.e., not only about knowledge of norms): The empiricist idea that cognition is possible without prior logical forms does not seem logically tenable; Rödl 2005, p. 16 et seq.

been said and what has been developed on the basis of it can be formulated in such a way that certain problems of classical discourse ethics do not arise.[14]

All this does not mean, for example, that self-interested preferences (or de facto national traditions) do not play an essential role for the actual enforcement, i.e., for the governance of climate protection and often (but not always) explain human behavior (Sect. 2.2.1). All that has been shown herewith is that no basic ethical-legal theory (e.g., with reference to sustainability) can be developed in this way. The basic intention of the cost-benefit analysis – everyone should be able to decide for themselves, and balancing is necessary – is, on the other hand, certainly shared by the present approach, as has already been made clear (Sect. 4.3). However, this does not apply to the theoretical formulation, the ignorance about sustainability, the plebiscitary orientation and the far-reaching promise of exactly specifiable balancing results.

In any case, what has been said now shows in more detail that normative ethics cannot be found by determining empirical preferences for altruism or the like.[15] Furthermore, ethics is not to be equated with a somehow non-anthropocentric line of thought that diffusely goes beyond mere monetary thinking – which would

[14]A whole series of conceivable objections against this derivation (1) of the possibility of reason and (on this, in a moment) (2) of dignity and impartiality as sole universal principles from reason are discussed in Ekardt 2019, Ch. 3.1. Rejected is, e.g., the argument from analytic philosophy that the question of the possibility of norms is on a different level than the other questions; likewise, the accusation of cultural imperialism is rejected (which, in short, could anyway only fit concepts that are far more detailed in content than the approach advocated here). Likewise, as seen (Sect. 4.3), only a normative space is determined, but mostly not precisely one correct normative answer, as it was envisioned by many from Plato to Dworkin in different ways. Rather, positions have already been taken on balancing necessities. The analysis also assumes no necessity of consensus-seeking as a property of language, no "obligation to debate" and no obligation to be rational. Furthermore, the present approach does not require a separate level of "ideal discourse." Moreover, the present approach explicitly leaves open the question whether language can also be private language and whether the individual or the collective has the "authority over language." Talking of a "category reason" (like in the present volume and in Ekardt 2019, Ch. 3.1) appeals less to language than to a particular entity. Thus, a list of good reasons is neither necessary nor intended (what further follows from this in terms of content will become apparent later in this section). Possibly some experience this as a relapse into a philosophy of consciousness (critical of such Böhler 1985, p. 29 et seq.), as does the distinction between definition and content pursued here. I do not see, however, how one should avoid that distinction. Even if someone claims to speak a private language, the structure of justification remains a structure of justification. Therefore, in my opinion, it does not matter whether "theories of meaning" are about what is meant, about what is understood, about meaning, or about whatever; see Liptow 2004.

[15]However, exactly this method is visible in TEEB 2009, where this method is significantly not called ethics, but political science.

amount to fixing ethics to an ecocentric approach (= granting animals and plants their own rights).[16] Although such an approach is advocated by some ethicists, it is (as has been discussed in detail elsewhere) more asserted than substantiated. This is the case, i.e., because it would ultimately not be possible to distinguish at all, without reference to humans, which parts of the almost incalculably complex natural world are worthy of protection (and which are not).[17]

An alternative ethics to the cost-benefit analysis is offered by the liberal-democratic basic principles already discussed in Chaps. 3 and 4. What is said there is ultimately justified by the interpretation of human dignity, the fundamental norm of liberal democratic constitutions as the necessary respect for the autonomy of the individual. That dignity is the ground of freedom is explicitly stated in liberal-democratic constitutions (in the Kantian tradition). But why should human dignity, as a necessary respect for autonomy, be compellingly justified? This brings us to the classical basic questions of a modern (heterodox) discourse ethics, which will be repeated here only very briefly. However, it would not change much the arguments against the cost-benefit-analysis, if not the following, but another justification-theoretical point of view would be considered applicable.

In a pluralistic world, people necessarily argue about normative issues. Even fundamentalists and autocrats inevitably do so at least occasionally. And they use human language to do so. But those who argue with reasons (i.e., rationally; i.e., with words like "because, since, therefore"), i.e., who formulate sentences "X is correct because Y" in normative questions, logically presuppose (1) the possibility of objectivity in normativity and (2) autonomy as normative core principle, whether they want it factually or not:[18]

---

[16] It becomes visible that exactly this is (incorrectly) assumed, e.g., in TEEB 2009. In contrast, Nestle 2012, p. 114 at least remarks that ethics does not simply merge into factual preferences; however, also there the noncognitivist approach is maintained to the effect that ethics is not to be justified, but only to be "disclosed" (and thus is obviously considered to be ultimately arbitrary); likewise TEEB 2010 declares values to be subjective without further ado, adding on p. 23 that a metric approach is also important for values, which is again empiricist. On the epistemological distinctions see above in Sect. 1.4.

[17] For more on this, see Schaller-Picard 2005; for other logical defects of the ecocentric approach (is-ought fallacy, etc.) see Ekardt 2019, Ch. 3.4. What is meant by economists is often covered up in English-language texts by the very vague term equity, which can equally mean justice, social distributive justice and equality. Exemplarily pars pro toto see TEEB 2010. In this respect too defensive (as a defender of cost-benefit analysis) with concessions to non-anthropocentric approaches Hansjürgens 2015, p. 1 (5 et seq.), if he means to accuse economic evaluation of a lack of consideration of "social" values. Indeed, if one assumes that the epistemological basis of cost-benefit analysis holds, it would already be unclear how one should elucidate such values other than by identifying individual preferences. And even if one doubts the theoretical basis of cost-benefit analyses (as in the present volume), just the convincing intentions of economic evaluations, namely their individualism, are given away.

[18] See in detail with further references Ekardt 2019, Ch. 3.1.

1. We logically presuppose (as seen) that normative questions can be decided at all with reasons and ergo objectively and not only subjectively preference-driven; otherwise, we contradict ourselves.
2. We further logically presuppose that the possible discourse partners deserve equal respect for their autonomy. For reasons are egalitarian and the opposite of violence and disparagement; and they are addressed to individuals with mental autonomy, for without autonomy one cannot examine reasons. No one could say "My thesis X and its reasons would be easily refuted by Mr. P, but you, dear Q, as a fool, should simply believe them." Nor would anyone be able to say, "After silencing P, we could finally convince ourselves that X is a good reason for Y." Thus, it is precisely contrary to the meaning of "reasons" to understand reasons as relative to the person to whom they are addressed; a reason convinces and can be understood by anyone. Someone who gives reasons in a conversation about justice (i.e., speaks sentences with "because, therefore, since") but then disputes the respect of the interlocutor, would ergo contradict what he himself logically presupposes. This is true regardless of whether he is aware of the implications of his reasoning or whether he thinks to discuss for mere persuasion purposes. For it is about strictly logical implications of our speaking (not, on the other hand, about our purely factual self-image, from which, taken by itself, nothing at all follows). For the purposes of the present volume, which is not primarily devoted to discourse ethics, it is irrelevant whether a (second) elenctic or transcendental argument is thereby presented, or whether a kind of alternativelessness without strictly logical derivation is stated in the sense of: Whatever the practice of reasoning implies, it fits better to liberal democracy than to any other normative conception, and at the same time one needs any norms, if one has proved before that there are reasonable norms in principle – for which then the same respect for autonomy is offered.

The rationally founded respect for autonomy as self-determination must now, however, apply precisely to the individual and thus be respect for individual autonomy: collectives as such are not possible partners in discourse at all. This is rather the individual arguing. This is the justification for the principle of respect for the autonomy of the individual (human dignity[19]). In addition, but hardly distinguishable from it, since the idea of equality is ultimately already inscribed in human dignity,

[19] This principle of human dignity is not itself a freedom/ fundamental/ human right. This principle is even not a norm tailored to solve concrete individual cases at all, neither legally nor ethically. Rather, human dignity is the ground of the spheres of freedom or human rights, rather than being a right itself; it thus directs the application of the other norms, in this case the various spheres of freedom of the citizens concerned, and provides autonomy as the guiding idea of the legal order. The "inviolability" of dignity and its character as a ground of rights, which is also visible in norms such as Art. 1 para. 2–3 of the German Basic Law and Art. 1 of the EU Charter of Fundamental Rights show that this makes sense not only philosophically but also in terms of legal interpretation; on the state of the debate in this regard, Ekardt 2019, Ch. 3.2.; Ekardt 2021, § 4 B.; Enders 1997; against this, for example, Böckenförde 2003.

the principle is thereby also justified that justice means independence from subjective perspectives (impartiality; whereby impartiality will not be dealt with in more detail here, since its message content is likely to coincide to a large extent with human dignity in the present reading[20]). From both principles, in turn, follows – not only legally (as mentioned), but also ethically – the right to freedom for all people, including (see Sect. 4.3) the further implications attached to it, such as the right to the elementary preconditions of freedom, the recognition also of other concerns that promote freedom, corresponding institutions of democracy based on the division of powers, and the rules of deliberation. And only this can be seen as justified; in the absence of a compelling justification, other principles cannot compete with it.[21] Therefore, equal self-determination, together with the principles and institutions presupposed, promoted and implied therein, is the criterion of justice.

These guarantees apply to all people, even if I never talk to them. For reasons in questions of justice (unlike statements in private or aesthetic questions) are addressed to everyone who could potentially refute them – which implies that I must recognize all people as to be respected, as soon as I speak in reasons at all sometimes in life; and everyone does. This again makes an example clear as a control consideration. Nobody could seriously say: "The absent Mr. P could refute my theses at any time – but you should believe them because of your stupidity." Whoever says such a thing would just have substantiated nothing. This is the further justification for an intertemporal (and cross-border) effect of fundamental rights suggested earlier. In the same way, it can be justified that normative derivations are binding not only in discourse, for instance, but also in action.[22]

It is also important that the principle of freedom is justified comprehensively here, i.e., not only freedom of speech, for example. For also property, free assembly, free choice of profession etc. etc. indirectly (sometimes more, sometimes less) promote autonomy and thus discourse. As already seen, this does not mean that these contents are not subject to balancing. Nevertheless, they are prima facie (legally and

[20] See again with further references Ekardt 2019, Ch. 3.1.

[21] A detailed proof of this cannot really be given in the present context, because various (ultimately infinitely many) other candidates for norms would have to be examined and it would have to be shown that these are either unfounded (because they derive from dogmatic settings, circular reasoning, or infinite recourses) or that they are ultimately consequences of the principle of freedom; for a number of conceivable approaches, this is precisely what is demonstrated in Ekardt 2021, § 3 B.-D.

[22] Habermas 1983, p. 96 et seq.; Ekardt 2019, Ch. 3.1. The openness of reason makes it inevitable to keep the further discourse about a question (or a new discourse about another question) open in principle. And whoever violates the principle of human dignity, even outside of a discourse, would limit the possibility of further discourse; for action (e.g., a law or even an individual act) does, after all, set the framework for any further discourse.

ethically) guaranteed (without differentiating between true and false freedom, such as the freedom to consume trash TV shows: Sect. 2.2.6).[23]

In practice, according to what has just been said, an ethical justification of environmental protection and sustainability succeeds by determining an obligation to sustainability and a contouring of the scope in this respect based on the fundamental principles of liberal-democratic constitutions and by ethically substantiating these foundations. This works without having to assume that ethics is necessarily able to provide an exact answer to individual questions of daily or political life. The basic assumption of the cost-benefit analysis, that it must take place in the absence of justifiable normative approaches, is therefore not convincing.

Cautious tendencies in the direction visible in all of this can occasionally be discerned in economic discourse as well. Thus, there are also efforts among economists (although not majoritarian) to carry out the cost-benefit analysis partly qualitatively instead of quantitatively.[24] It is then conceded that not everything could be put into a numerical form, such as the concrete weight of a human life. Structurally, this seems to be a step towards more differentiated approaches. Nevertheless, central problems remain. The orientation towards purely factual preferences with all the fundamental problems described above remains. With regard to the problems of application, however, a far-reaching reduction occurs; the handling of complexity and uncertainty is then simply the same as in the legal or ethical balancing approach described above. However, the empiricist approach becomes unfaithful to itself here, in that it virtually presupposes that people do not know what they want and are not able to articulate their "true interests" (which can hardly exist under empiricist conditions).

Representatives of ecological economics, which is critical of neoclassicism, go even further, advocating a value objectivism in ecological terms, which, however, is not further substantiated, does not develop any concrete balancing rules and, moreover, deprives itself of a basis by its lack of a clear commitment to normative

[23] The focus on a universalistic principle of freedom is also not slightly bizarre, in that the principles implied by the discourse at the same time conclusively determine the possible content of discourses of justice, and the entire ethics and legal order is thus ultimately brought into terms of freedoms and preconditions of freedom as well as conditions conducive to freedom and their exceedingly numerous (also institutional) implications. Although this is the case, it is nevertheless not wrong, because (a) the discourses of the good life are left with an arbitrary leeway, which do not experience any normative specification here (because such a specification is not even possible), because (b) also in discourses of justice the discourse conditions do not have to be thematized as discourse conditions, because the discourse conditions sufficiently depict the possible complexity of discourses of justice, because (c) great leeway for balancing remains, and because (d) only in this way the principle of autonomy can be preserved, and so on. See Ekardt 2021, § 3 F. and § 4 F. I.; Ekardt 2019, Ch. 3.1.

[24] See, for example, Lienhoop and Hansjürgens 2010/2011, p. 229 et seq. (with some incorrect renditions of my position); Spash 2007, p. 690 et seq.; Brouwer et al. 2013; MacMillan et al. 2006, p. 299 et seq.; Jakob and Steckel 2016; see also Gawel 2001, p. 9 et seq. and 43 et seq.; more interested in a more reflective monetization and less in complementing it: Nestle 2012, p. 114 et seq.; Hirschfeld, Costs, p. 277 et seq.

reason.[25] However, this would gain little in terms of justification compared to the cost-benefit analysis.

5.2 Economic Ethics: Modified Empiricist/Utilitarian Ethics Instead of Cost-Benefit Analysis as Way Out?

The assessment thus developed of the justification-theoretical frictions of cost-benefit analysis is also relevant to the climate debate insofar as empiricism is also widespread in explicitly ethical treatises on climate issues.[26] This is not always obvious, but it becomes clear when, in particular, factual moral intuitions are repeatedly referred to.[27] This is often combined with a recourse to John Rawls, whose basic principles of equal respect and impartiality are not justified in an elenctic way (as in Sect. 5.1 above), but with their factual recognition.[28] As has already been made clear (Sects. 2.2.3, 2.2.4, and 5.1), such approaches are often quite at odds with cost-benefit analysis, because they criticize, for example, the constant balancing or quantifying or the allowing of arbitrary (especially self-interested) factual preferences. As has already been pointed out, there are nevertheless analogies to the criticism of cost-benefit analysis with regard to the justification-theoretical frictions (Sect. 5.1). In the present chapter, this will not be pursued in full depth, but at least a brief look at approaches with close affinities to cost-benefit analysis seems necessary, as they are pursued particularly by economists. These approaches typically also assume (mostly self-interested) factual preferences, but without a quantification element. If one thinks in this direction, similar and partly even further frictions arise, as will be briefly shown here.

It is not possible here to take up the entire, centuries-old discussion of utilitarian approaches in all their diversity. It is obvious that, in addition to the theoretical frictions of linking to factual preferences, the problem of the legal compatibility of plebiscitary preference approaches applies to a utilitarian theory just as it does to cost-benefit analysis. In contrast, it has already been made clear that individualism and balancing as such do not deserve criticism in both approaches (Sects. 2.2.3 and 2.2.7). All this remains the case even if, instead of money, one were to try to

[25] See in detail on this with further references von Egan-Krieger 2014, p. 230 et seq.

[26] As an example Jamieson 2014, p. 115 et seq.; Moellendorf 2014, p. 90 et seq.; Gesang 2011, p. 78 et seq.; Gesang 2000; Gesang 2016.

[27] See for instance Lübbe 2015; Lübbe 2016, p. 325 et seq.; Lübbe 1998; Betzler and Schroth 2014, p. 279 et seq.

[28] Rawls 1971; exemplary for the adaptation Jamieson 2014; Moellendorf 2014; Schumann 2005, p. 109 et seq.; Hampicke 2005, p. 475 et seq.; Crane and Matten 2010, p. 45 et seq. and 235 et seq. Despite the explicit reference to factual intuitions, Breyer and Kolmar 2014, p. 50 mean that Rawls thereby only wants to follow Kant, who indeed speaks of the idea of impartiality respectively the categorical imperative as a fact of reason, whereby his approach in practical philosophy nevertheless appears to be clearly distinguishable from other empiricist approaches; on Rawls in more detail Ekardt 2021, § 3 C.

quantify happiness or preferences per se,[29] for example, in the utilitarian tradition, in which case it would be even less clear what exactly one has to measure, as has also been repeatedly noted in economic research[30] (the specific problems of empirical and normative happiness research have been the subject of elsewhere[31]). It also remains problematic with other preference approaches beyond cost-benefit analysis when there is again a mixing of empirical presentations of the human image and behavioral research on the one hand and normative ethics on the other.[32] The analogous problem has already been examined above for the relationship between homo oeconomicus and cost-benefit analysis (Sect. 2.2.1).

However, one particular approach deserves closer attention. As has already been made clear, cost-benefit analysis could also be called economic ethics. There is now a related approach that explicitly calls itself like that. This needs to be discussed in more detail now. In this context, the empiricist basis of neoclassical economics becomes even clearer when, in economic ethics, even supposed criticism of efficiency ultimately leads back into utilitarian or, in any case, again empiricist models.[33] Important representatives of this ethics are Karl Homann, Andreas Suchanek and Christoph Lütge.[34] For those familiar with law, the initial demands for more realism in the ethical debate are relatively familiar in the sense that, in view of the unmanageable diversity of ethical debates, ethics must be sought in the concretely binding statements of law, that ethics must start at the social level instead of the individual level, and that solidarity presupposes that distributional mass is generated at all. The further point that people are allegedly purely factually self-interested and that the framework order must therefore be built in such a way that it serves the self-interest of everyone simply repeats the already known Hobbesian basic ideas (Sect. 2.1.1). In the end, competition allegedly benefits everyone, state control must be limited, and the free play of forces, albeit according to rules, is the solution to environmental problems from the viewpoint of economic ethics.

---

[29] See on this Singer 1979; on the background also Hare 1952; Fehige and Meggle 1995.

[30] Arrow et al. 1997; Susnjar 2010, p. 215 et seq.; Breyer and Kolmar 2014, p. 50 et seq.; Scholz 2011, p. 192 et seq.

[31] See with further references Ekardt 2019, Ch. 2.6 and 3.4.

[32] As an example Haller 2012, p. 39 et seq.; Crane and Matten 2010, p. 45 et seq. und 235 et seq.; Müller-Christ 2014, p. 100 et seq. und 225 et seq.

[33] Visible, for example, in the discussion on efficiency and consensus in Lenger and Goldschmidt 2012, p. 53 et seq. and Wilhelmi 2012, p. 95 et seq.

[34] Homann and Lütge 2013, p. 17 et seq.; Lütge 2012, p. 89 et seq.; Suchanek 2000, p. 101 et seq.; Lütge 2004, p. 117 et seq. – Sen 2009, p. 119, 121 et seq. and passim also attempts a quantifying normative social theory that is not reduced to purely monetary preferences. In doing so, he correctly demands that global justice, balancing, and the real enforcement of justice have a place in justice theory. However, his idea that a general theory of justice is not necessary as a normative criterion and that such a theory is not helpful at all to be able to say something about real (= non-ideal) conditions is unconvincing – see Sect. 5.1. Strange is his use of the word transcendental (he seems to mean rather transcendent) and the very imprecise ideas about against whom human rights work at all.

However, it remains unclear according to which normative standards it should be determined how the framework should be designed, i.e., whether in the end any legal order should be recognized – and how this exactly relates to starting from factual preferences. Moreover, the normative justification and descriptive explanation of behavior are again mixed up (and a purely competitive-egoistic motivational situation of humans is descriptively inaccurately assumed; for the necessary differentiations, see Sect. 2.2.1). Furthermore, economic ethics again suffers from the fading out of the intertemporal and global-border problem situation just as cost-benefit analysis.[35] Furthermore, competition (to address this very complex issue at least briefly) will by no means always benefit everyone[36]: Free world trade, for example, is sometimes of little use to developing countries today, because their still underdeveloped industries are then simply outcompeted by Western companies. This is precisely why even countries such as Germany relied on tariff barriers and protectionism during their industrialization phase in the nineteenth century and not on free competition. Because they knew that competition by no means always helps the weak. The idea that competition driven by self-interest ultimately benefits everyone because of its wealth-creating power is flawed in another respect: For ecological reasons, ever increasing prosperity is not an option, first in the industrialized countries, but in the distant future it will no longer be an option in other countries either. Problems such as climate change cannot be solved by exclusively technical means (Sect. 1.3), which, as seen, may mean saying goodbye to the growth society and thus to ever more economic gains (Sect. 3.4). Nevertheless, the market principle can also promote such far-reaching environmental protection. If, for example, resources are made more expensive by levies, competition for better environmental technology is stimulated in a competitive market economy (Sect. 2.2.8). At the same time, however, there is also an incentive to consume less. This is then precisely not the continuation of an eternal growth society as insinuated in economic ethics.

The problems become even greater if economic ethics is not read as a normative theory of politics, but as a recommendation to individual companies. Regarding the thoughts and frictions in this respect, reference can again be made to an earlier detailed treatment[37]: Many hope that consumers and companies will voluntarily step into the breach in favor of climate protection and other social goals (in the sense of corporate social responsibility/ CSR[38]). If the public authorities make such self-regulation more likely through limited frameworks such as information obligations

[35] Röttgers 2011a as well as Röttgers 2011b and (somewhat more friendly) Hemel 2011 also attest a mixture of a lack of concreteness, a lack of practicability and therefore a lack of relevance of current business ethics.

[36] On the discussion of opportunities and limits of competition Ekardt 2019, Ch. 4.2 with further references. And by no means does legally compliant behavior always and consistently promote entrepreneurial profit. Exemplary for such shortcomings: Girrbach 2010, p. 84 et seq.; Aufderheide 2015, p. 57 et seq.

[37] See with further references Ekardt 2019, Ch. 4.2; Ekardt 2021, § 6 B.

[38] Exemplary for the discussion and for the ignoring of the factors mentioned in the body text below Stelzer 2009, p. 11 et seq.; Austmann 2009, p. 30 et seq.; Wieland et al. 2009, p. 67 et seq.; Crane and Matten 2010, p. 45 et seq. and 235 et seq.; more closely related to discourse ethics, the anthology by Bausch 2008; a very good and balanced account of the debate is provided by Zimpelmann and Zöckler 2008, p. 119 et seq.

for companies, a solution to the problem (also) of climate protection could therefore allegedly succeed[39]). More broadly, the trend toward more competition or, more generally, toward more proceduralization, which has been growing for some time, is about the following.[40] The issue is what form of governance and justice realization is the effective and just one under the conditions of a globalized world with globalized markets and increasingly complex technology, where the old nation-state encounters seemingly increasing difficulties in enforcing its goals. But it is doubtful how the real and not just occasional problem of ethics and law – the balancing of conflicting freedoms or conflicting concerns – can be meaningfully solved at the level of a behavioral appeal to individual citizens and companies. If, for example, resolute climate protection should demand not only marginal changes but a rethinking of the entire Western consumption model: What then should the appeal to individual companies be? To stop producing large cars, but only bicycles and a few 1-liter cars, then to go bankrupt with them and let other companies produce the cars instead? And how much do I have to contribute as an individual actor in relation to others? These examples and questions underline and illustrate the finding, which has been deepened elsewhere, that the resolution of conflicts of freedom and the responsibility for the enforcement of fundamental rights do not have to lie with citizens and companies, but with national and transnational public powers.

Moreover, the problematic empirical background assumption behind the focus on corporate responsibility, consumer democracy, etc. is ultimately that companies and consumers could continue their consumption, production and growth style essentially unchanged, with the only proviso that the individual products would have to become successively better in terms of their environmental balance. After all, the driving force behind this is the interest in buying and selling more products and more services. As has already been pointed out, however, such a strategy of, e.g., technically better cars, but more and more cars cannot bring about the necessary reduction in emissions and resource extraction; purely technical change is not enough for sustainability, and even this change does not get off the ground fast enough by itself. To put it succinctly: More and more consumer electronics, long-distance travel, exotic foods, large apartments and car journeys will not become ecologically compatible by subsequently greening them in part through technical improvements that can be initiated by business. On the contrary, the overall balance will foreseeably fall further short of the climate and biodiversity targets mentioned above.[41] At the same time, the problems described in this paragraph make it clear that setting CSR and consumers, for example, is not only an (unconvincing) individual ethics, but at the same time also a (only foreseeably less effective)

[39] For a summary on the following: Salzborn 2011, p. 43 et seq. und 121 et seq.

[40] See on these tendencies also Böhm 1996, p. 6 et seq.; Getliffe 2002; Lee and Abbot 2003; Ekardt 2010; classical Luhmann 1997; Habermas 1992, p. 316 et seq.

[41] The discourse on corporations as possible moral actors is a misunderstanding in all this, since rights and duties correlate very well here, and this cannot be negated by analyses of the capacity to suffer. Therefore, Neuhäuser 2011, p. 90 et seq. cannot be followed.

governance approach that competes with economic instruments and regulatory law, for example.[42]

The frictions of economic ethics described in this way also occur to a large extent analogously in approaches to business ethics, which are at least partly formulated by economists. Here, too, questions of behavioral science are mixed up with the question of what companies and consumers are supposed to do (is-ought fallacy).[43] Partly, it is also argued from factual enforcement that ethics must aim at attitude rather than just values.[44] Furthermore, often no substantive normative statement is made at all, but rather, beyond generally well-sounding formulas about the responsibility of companies. Or simply an empirical field is described, such as the factual actions of a company, accompanied by the statement, not further justified normatively, that this is to be welcomed in some way. Moreover, even where the necessity of an independent normative argumentation is recognized in principle, such an argumentation is not developed in more detail. Instead, it is relatively quickly assumed that, due to the social interaction of people, some normative obligations exist, for example, between companies, consumers and affected third parties. The fact that this is both a naturalistic fallacy and a mere assertion without a justified normative basic theory is not discussed. It is also problematic at this point if a habitus, i.e., an inner attitude, is regarded as sufficient for the preformation of complex decisions which, after all, require balancing rules.[45] (In addition, the problems arise here that one can hardly plan such a habitus and that also the idea, usually connected with such demands, that there is a decay of morals or the like, appears too undifferentiated and rather as a classical theory of decay – ultimately belonging to the philosophy of history.[46]) Finally, the real impact of supposedly corporate/ consumer ethical behavior is largely ignored; for example, the fact that efforts to date in this regard have made little difference to the devastating climate and resource balance of Western countries and what a dramatically large need for action there actually is, receives little attention.[47]

[42] See in detail (and critical) with further references Ekardt 2019, Ch. 4.2.

[43] As an example for the following: Beschorner et al. 2008, p. 9 et seq. passim; Müller-Christ 2014, p. 100 et seq. and 225 et seq.; Schmidt et al. 2008, p. 9 et seq.; Schaltegger et al. 2013.

[44] See on this Manemann 2012, p. 11 et seq.; critical also Sen 2009.

[45] See in this direction also Manemann 2012, p. 11 et seq.

[46] Critical of philosophical approaches to history, whether designed as a theory of decay or as a theory of progress: Ekardt 2017a, b, Ch. 8.

[47] Examples of other approaches to economic ethics are, e.g., Pfriem 2007 and Pfriem 2016 (with an extremely broad concept of "cultural" and "education"; a mixing of normative justification and real enforcement as well as of subjectivity with normativity; an equation of cultural and subjective; in contrast, the normative orientation of cost-benefit analysis as well as more recent developments in discourse ethics do not come into view there).

5.3 Analogous Problems of Crypto-Natural Scientific Risk-Benefit Analyses and Sustainability Indicators

Preference theory is the basis for how much environmental protection the respective economists consider objectively correct or efficient.[48] At the same time, the cost-benefit analysis admits – and this is just true – that it is about human judgments and not about the fact that the value of an ecosystem or the global climate as a natural fact lies in these things themselves. This is very important, because this human and just not natural scientific side is very often overlooked in quite different contexts, each of which is highly controversial in itself, without, however, mostly the actual problem (what normativity actually means) being perceived.

The problems are rather bigger, if now the preferences are given up as a point of reference and – still following the empiricist ideal of knowledge – the balancing material respectively advantages and disadvantages are really to be generated from the things themselves.[49] This is done, for example, under terms such as risk-benefit analyses, which, however, because they cannot justify any normative criterion, appear even less plausible than cost-benefit analyses. For instance, natural science data on possible climate developments alone cannot be used to determine whether they are beneficial or harmful. One attempt to solve this problem is the so-called sustainability indicators. Countries and companies are constantly striving for indicators and a measurability of sustainability to make sustainability visible in a simplified form through a few easily quantifiable aspects selected from the multitude of relevant factors: such as greenhouse gas emissions, land consumption, energy consumption per capita, the share of renewable energies in electricity generation or the water quality of certain large rivers.[50] Real measurability is often still sought to the extent that all these things are to be offset against each other. In this way, certain development tendencies, their advantages and disadvantages, and one's own (real or supposed) successes are to be visualized and made understandable for a wider audience. In the case of numerical indicators, the state, the tendency in absolute terms, the tendency relative to the past and the tendency relative to a target value can be assessed.[51] Of course, it is also relevant whether the measurement is correct at all.[52] However, indicators are skewed in many respects. The permanent and global viability of economic and living systems is not reflected, for example, when a company decides to produce 5-liter instead of 8-liter cars in the future. Another problem with

[48] So too, despite their opposition in other points, Stern 2009, Ch. 3, 5 and Nordhaus 2008, p. 38 et seq. and 59 et seq.

[49] Critical Schaller-Picard 2005; Meyer 2003, p. 150 et seq.; Rogall 2013, p. 73 et seq.; Baumgärtner 2012; Scholz 2011; on the following see already Ekardt 2019, Ch. 1.5.

[50] UBA 1997, p. 318 et seq. (also on the inexistence of meaningful social indicators).

[51] See Hulliger and Lussmann 2009.

[52] The following are proposed as "quality criteria" for sustainability indicators: a representative depiction of the situation, causes of change, societal response actions, simple structure, easy interpretability, clarity, international comparability, possibility of evaluation by reference values – see Bielig 2003, p. 2.

indicator and measurement approaches is that seemingly precise individual factors suggest a degree of accuracy that does not actually exist, because the factors are ultimately selected arbitrarily. In normative terms, sustainability indicators do not provide any founded justification and therefore do not provide a criterion for determining the advantages and disadvantages of sustainability decisions.

The continued attempts to somehow deduce normative decisions from purely factual findings on sustainability are also related to the problem that the distinction descriptive versus normative is often blurred behind supposedly homogenizing terms such as "professionally appropriate" or "normal." When, for example, one speaks of (a) the appropriate state of nature and ecosystems, (b) or of an appropriate physical and mental state of human beings (as opposed to illness), (c) of the well-being of children, for example of a care situation that is conducive to young children (some argue for the crèche, others for the mother) or (d) normal sexual preferences (in contrast to paraphilias, which are understood as ultimately pathological and which used to include homosexuality, but today are more likely to include pedophilia, but many also include parthenophilia), then in this exemplary spectrum descriptive and normative elements are mixed on several levels. In the constellations just mentioned, the question is initially obscured as to which parts of these problems are empirical and which parts are normative. It imposes normative questions and not empirical questions whether a certain state of nature is desirable, whether a long healthy life is worth striving for, whether children should do well with great/ medium/ no consideration also for the concerns of their parents, and whether sexual activities towards same-sex partners, significantly younger people or children are to be criticized. Thus, they cannot be answered by natural scientists. There is no "normal" state of nature, but a huge number of conceivable states, and these often reveal themselves in the course of centuries and millennia. A question of fact, however, is whether certain ecosystems and ecosystem services can persist when certain impairments occur. It is also a question of fact in the other example that certain human bodily states designated as disease and thus as "not normal" shorten life expectancy or are otherwise not desired by the person concerned. This means in other words that the presumed self-interest of the respective person is threatened (or, when it comes to mental states endangering others, as in the case of pedophilia, the interest of others involved).

The above criticism, on the other hand, does not refer to attempts to depict sustainability-related interdependencies in a purely descriptive manner. Attempts to depict the complex interdependencies of environmental problems and the manifold implications of, for example, a product or an energy source in so-called life cycle assessments appear to be quite reasonable. In doing so, an attempt is made to comprehensively describe the various impact variables empirically without evaluation. This can also be done in such a way that everything is considered from one point of view, for example, with regard to the respective land use or the respective energy consumption. However, the ignorance-related or complexity-related limits of knowledge, which are familiar from the application problems of cost-benefit analysis, also play a role here. It must also be stated that the area reference, for example, and also any other attempt to reduce all sustainability-related impacts to one measure, can only inadequately represent the totality of ecological issues. Ultimately, there is also no need to homogenize the complex set of facts regarding the totality of ecological implications of a practice (such as conventional agriculture) into a countable quantity. When economists, for example, try to describe the state of ecosystems comprehensively, usually under the heading of natural capital accounting, they are therefore well advised to carefully juxtapose the ascertainable physical quantities without necessarily aiming for a conversion into uniform quantities.[53]

Repetition Questions

1. What – especially logical – fundamental problems arise from the basic empiricist approach to cost-benefit analysis? (Sect. 5.1)
2. What is meant by a performative (self-)contradiction? (Sect. 5.1)
3. How can the central liberal-democratic basic concepts of reason, dignity, freedom and democracy be linked instead? (Sect. 5.1)
4. What problems arise when an economic ethics or utilitarian approach is adopted instead of a cost-benefit analysis? (Sect. 5.2)
5. To what extent do the frictions of cost-benefit analysis recur in risk theory and with regard to sustainability indicators? (Sect. 5.3)

References[54]

Alexy R (1995) Recht, Vernunft, Diskurs. Frankfurt a.M

Arrow K et al (1997) Is there a role for benefit-cost analysis in environmental, health and safety regulation? Environ Dev Econ 272:196 et seq

Aufderheide D (2015) Zur Debatte um die gesellschaftliche Verantwortung von Unternehmen. Ist die CSR-Bewegung Teil der Lösung oder Teil des Problems? In: Aufderheide D, Dabrowski M (eds) Markt und Verantwortung. Wirtschaftsethische und moralökonomische Perspektiven. Duncker & Humblot, Berlin, p 57 et seq

Austmann H (2009) Corporate Social Responsibility und nachhaltige Entwicklung, Hamburg

Baumgärtner S (2012) Normative Begründung der Nachhaltigkeitsökonomik. In: Studierendeninitiative Greening the University (ed) Wissenschaft für nachhaltige Entwicklung! Multiperspektivische Beiträge zu einer verantwortungsbewussten Wissenschaft. Metropolis-Verlag, Tübingen, p 273 et seq

Bausch T (ed) (2008) Normativität und Anwendungsbedingungen einer Wirtschafts- und Unternehmensethik in marktwirtschaftlichen Wettbewerbsstrukturen. Lit, Münster

Beschorner T, Schmidt M, Vorbohle K, Schank C (eds) (2008) Zur Verantwortung von Unternehmen und Konsumenten. Rainer Hampp Verlag, Munich et al

Betzler M, Schroth J (2014) Konsequentialisierung – Königsweg oder Sackgasse für den Konsequentialismus? Z Philos Forsch 68:279 et seq

[53] See in detail with further references Ekardt 2017a, b, p. 9 et seq.

[54] In accordance with legal practice, parliamentary, governmental and EU Commission documents as well as laws and judgments are not listed in the bibliography, as they can be found unequivocally based on the reference given in the continuous text or via the general search engines. The last access date for all Internet sources is 31.10.2021.

Bielig A (2003) Messung von Nachhaltigkeit durch Nachhaltigkeitsindikatoren. Technische Universität Ilmenau, Ilmenau
Böckenförde E-W (2003) Menschenwürde als normatives Prinzip. Juristenzeitung:809 et seq
Böhler D (1985) Rekonstruktive Pragmatik. Von der Bewusstseinsphilosophie zur Kommunikationsreflexion – Neubegründung der praktischen Wissenschaften und Philosophie. Suhrkamp, Frankfurt a.M
Böhm M (1996) Der Normmensch. Materielle und prozedurale Aspekte des Schutzes der menschlichen Gesundheit vor Umweltschadstoffen. Tübingen
Breyer F, Kolmar M (2014) Grundlagen der Wirtschaftspolitik, 4th edn. Mohr Siebeck, Tübingen
Brouwer R et al (2013) A synthesis of approaches to assess and value ecosystem services in the EU in the context of TEEB. http://ec.europa.eu/environment/nature/biodiversity/economics/pdf/EU%20Valuation.pdf
Crane A, Matten D (2010) Business ethics. Managing corporate citizenship and sustainability in the age of globalization, 3rd edn. Oxford University Press, Oxford
Dilger A (2006) Ökonomik versus Diskursethik. Thesen zu Felix Ekardt. Zeitschrift für Umweltpolitik und Umweltrecht 29:383 et seq
Eidenmüller H (2005) Effizienz als Rechtsprinzip. Möglichkeiten und Grenzen der ökonomischen Analyse des Rechts, 3rd edn. Mohr Siebeck, Tübingen
Ekardt F (2006) Ökonomik versus Diskursethik in der Umweltpolitik: Antikritische Bemerkungen zu Alexander Dilger. Zeitschrift für Umweltpolitik und Umweltrecht 29:399 et seq
Ekardt F (2010) Information, Partizipation, Rechtsschutz. Prozeduralisierung von Gerechtigkeit und Steuerung in der Europäischen Union, 2nd edn. Münster
Ekardt F (2017a) Natural capital accounting and the mobilization of financial resources for biodiversity, Analyses for the German Federal Government. Leipzig. http://felix-ekardt.eu/files/texts/Biodiv-Evaluation-Instruments.pdf
Ekardt F (2017b) Wir können uns ändern. Gesellschaftlicher Wandel jenseits von Kapitalismuskritik und Revolution. München
Ekardt F (2019) Sustainability. Transformation, governance, ethics, law. Springer, Dordrecht
Ekardt F (2021) Theorie der Nachhaltigkeit. Ethische, rechtliche, politische und transformative Zugänge – am Beispiel von Klimawandel, Ressourcenknappheit und Welthandel, 4th edn. (= 3rd ed. of the new edition). Baden-Baden
Enders C (1997) Die Menschenwürde in der Verfassungsordnung. Zur Dogmatik des Art. 1 GG. Tübingen
Fehige C, Meggle G (eds) (1995) Zum moralischen Denken, 2 Bde. Frankfurt a.M
Fromm O (1997) Möglichkeiten und Grenzen einer ökonomischen Bewertung des Ökosystems Boden. P. Lang, Frankfurt a.M
Gawel E (2001) Ökonomische Effizienzforderungen und ihre juristische Rezeption. Ein problemstrukturierender Überblick. In: Gawel E (ed) Effizienz im Umweltrecht. Grundsatzfragen einer wirtschaftlichen Umweltnutzung aus rechts-, wirtschafts- und sozialwissenschaftlicher Sicht. Nomos, Baden-Baden, p 9 et seq
Gayer T, Viscusi K (2014) Determining the proper scope of climate change benefits. Manuscript
Gesang B (2000) Kritik des Partikularismus. Über partikularistische Einwände gegen den Universalismus und den Generalismus in der Ethik. Paderborn
Gesang B (2011) Klimaethik. Berlin
Gesang B (2016) Wirtschaftsethik und Menschenrechte. UTB, Tübingen
Getliffe K (2002) Proceduralisation and the Aarhus convention. Environ Law Rev 4:101 et seq
Girrbach P (2010) Implementierung von Corporate Responsibility. Entwicklung eines nachhaltigkeitsorientierten Transformationsmodells für die unternehmerische Praxis. Mensch-und-Buch-Verl, Berlin
Habermas J (1983) Moralbewusstsein und kommunikatives Handeln. Frankfurt a.M
Habermas J (1992) Faktizität und Geltung. Frankfurt a.M
Haller C (2012) Menschenbild und Wirtschaft. Eine philosophische Kritik und Erweiterung des Homo oeconomicus. Tectum Wissenschaftsverlag, Marburg

Hampicke U (2005) Gedanken zu minimaler Gerechtigkeit. In: Beschorner T, Eger T (eds) Das Ethische in der Ökonomie. Festschrift für Hans Nutzinger. Metropolis-Verl, Marburg, p 475 et seq
Hampicke U (2011) Climate change economics and discounted utilitarianism. Ecol Econ 72:45 et seq
Hanley N, Barbier E (2009) Pricing nature. Cost-benefit analysis and environmental policy. Elgar, Cheltenham
Hansjürgens B (2015) Zur Neuen Ökonomie der Natur. Kritik und Gegenkritik. Wirtschaftsdienst 95(4):1 et seq
Hare R (1952) The language of morals. Clarendon Press, Oxford
Harsanyi JC (1978) Rule utilitarianism and decision theory. In: Gottinger HW, Leinfellner W (eds) Decision theory and social ethics. Issues in social choice. D. Reidel, Dordrecht et al, p 3 et seq
Hemel U (2011) Vom Sinn und Nutzen einer philosophisch fundierten Wirtschaftsanthropologie. fiph J 18:22 et seq
Homann K, Lütge C (2013) Einführung in die Wirtschafsethik, 3rd edn. Münster
Hulliger B, Lussmann D (2009) Bewertung von Nachhaltigkeits- und Umwelt-Indikatoren. Hochsch. für Wirtschaft, Olten
Illies C (2003) The grounds of ethical judgement – new transcendental arguments in moral philosophy. Clarendon Press, Oxford
Jakob M, Steckel JC (2016) Implications of climate change mitigation for sustainable development. Environ Res Lett:104010
Jamieson D (2014) Reason in a dark time. Why the struggle against climate change failed – and what it means for our future. Oxford University Press, Oxford
Kuhlmann W (1985) Reflexive Letztbegründung. Freiburg et al
Lee M, Abbot C (2003) The usual suspects? Public participation under the Aarhus convention. Mod Law Rev 66:80 et seq
Lenger A, Goldschmidt N (2012) Effizienz oder Konsens? Einige prinzipielle Überlegungen zum Verhältnis zweier Grundkategorien ökonomischen Denkens. In: Aufderheide D, Dabrowski M (eds) Effizienz oder Glück? Wirtschaftsethische und moralökonomische Perspektiven der Kritik an ökonomischen Erfolgsfaktoren. Duncker & Humblot, Berlin, p 53 et seq
Lienhoop N, Hansjürgens B (2010/2011) Vom Nutzen der ökonomischen Bewertung in der Umweltpolitik. GAIA:255 et seq. und 229 et seq
Liptow J (2004) Regel und Interpretation. Eine Untersuchung zur sozialen Struktur sprachlicher Praxis. Velbrück Wissenschaft, Weilerswist
Lübbe W (1998) Verantwortung in komplexen kulturellen Prozessen. Breisgau Alber, Freiburg et al
Lübbe W (2015) Nonaggregationismus. Grundlagen einer Allokationsethik. Münster
Lübbe W (2016) Handlungen, Handlungskonsequenzen und das Vollständigkeitsaxiom. Ein handlungstheoretischer Kommentar zum entscheidungstheoretischen Konsequentialismus. Z Philos Forsch 70:325 et seq
Luhmann N (1997) Legitimation durch Verfahren, 4th edn. Frankfurt a.M
Lütge C (2004) Ordnungsethik – naturalistisch konzipiert. In: Lütge C, Vollmer G (eds) Fakten statt Normen? Zur Rolle einzelwissenschaftlicher Argumente in einer naturalistischen Ethik, Baden-Baden, Nomos, p 117 et seq
Lütge C (2012) Wirtschaftsethik ohne Illusionen. Ordnungstheoretische Reflexionen. Mohr Siebeck, Tübingen
MacMillan D, Hanley N, Lienhoop N (2006) Contingent valuation: environmental polling or preference engine? Ecol Econ 60:299 et seq
Manemann J (2012) Unterwegs zu einer habituellen Unternehmensethik. In: Hemel U, Fritzsche A, Manemann J (eds) Habituelle Unternehmensethik. Von der Ethik zum Ethos. Nomos, Baden-Baden, p 11 et seq
Meyer M (2003) Überlegungen zur Rationalität institutionenökonomischer Modelle. In: van Aaken A, Schmidt-Lübbert S (eds) Beiträge zur ökonomischen Theorie im Öffentlichen Recht. Deutscher Universitätsverlag, Wiesbaden, p 149 et seq

Moellendorf D (2014) The moral challenge of dangerous climate change. Values, poverty, and policy. Cambridge University Press, Cambridge, MA
Müller-Christ G (2014) Nachhaltiges Management. Einführung in Ressourcenorientierung und widersprüchliche Managementrationalitäten, 2nd edn. Baden-Baden, Nomos
Nestle I (2012) The costs of climate change in the agricultural sector. A comparison of two calculation approaches, Dissertation, Flensburg
Neuhäuser C (2011) Unternehmen als moralische Akteure. Suhrkamp, Frankfurt a.M
Nordhaus W (2008) A question of balance. Weighing the options on global warming policies. Yale University Press, New Haven et al
Ott K, Döring R (2004) Theorie und Praxis starker Nachhaltigkeit. Metropolis-Verl, Marburg
Pearce D, Atkinson G, Mourato S (2006) Cost-benefit analysis and the environment. Recent development. Paris
Pfriem R (2007) Unsere mögliche Moral heißt kulturelle Bildung. Unternehmensethik für das 21. Jahrhundert. Metropolis-Verl, Marburg
Pfriem R (2016) Ökonomie als Gemengelage kultureller Praktiken. Metropolis-Verlag, Marburg
Pissarskoi E (2014) Gesellschaftliche Wohlfahrt und Klimawandel. Umgang mit normativen Annahmen und Ungewissheiten bei der klimaökonomischen Politikberatung. Oekom-Verl, Munich
Rawls J (1971) A theory of justice. Cambridge University Press, Cambridge, MA
Rödl S (2005) Kategorien des Zeitlichen. Eine Untersuchung der Formen des endlichen Verstandes. Suhrkamp, Frankfurt a.M
Rogall H (2012) Nachhaltige Ökonomie. Ökonomische Theorie und Praxis einer Nachhaltigen Entwicklung, 2nd edn. Marburg (3rd ed. angekündigt für 2018)
Rogall H (2013) Volkswirtschaftslehre für Sozialwissenschaftler. Einführung in eine zukunftsfähige Wirtschaftslehre, 2nd edn. Springer, Wiesbaden
Röttgers K (2011a) Warum Wirtschaftsphilosophie? FIPH J 18:6 et seq
Röttgers K (2011b) Wirtschaftsethik, Wirtschaftsmoral und die Aufgaben der Wirtschaftsphilosophie. In: Busche H (ed) Philosophische Aspekte der Ökonomie. Königshausen und Neumann, Würzburg, p 39 et seq
Salzborn N (2011) Das umweltrechtliche Kooperationsprinzip auf unionaler Ebene. Nomos, Baden-Baden
Schaller-Picard N (2005) Zum Verhältnis von „Natur" und „Schöpfung". Eine umweltethische Studie in naturwissenschaftlicher und schöpfungshermeneutischer Perspektive. Stuttgart
Schaltegger S, Hörisch J, Herzig C (2013) Methodische Operationalisierung unternehmerischer Nachhaltigkeit. Zeitschrift für Umweltpolitik und Umweltrecht 36:87 et seq
Schmidt M, Beschorner T, Vorbohle K, Schank C (eds) (2008) Corporate social responsibility und corporate citizenship. Munich et al
Scholz R (2011) Environmental literacy in science and society. From knowledge to decisions. Cambridge University Press, Cambridge
Schumann O (2005) Wirtschaftsethik und Politische Ökonomie in theoriegeschichtlicher Perspektive. In: Beschorner T, Eger T (eds) Das Ethische in der Ökonomie. Festschrift für Hans Nutzinger. Metropolis-Verl, Marburg, p 109 et seq
Sen A (2009) The idea of justice. Harvard University Press
Singer P (1979) Practical ethics. Cambridge University Press, Cambridge
Smart JJC (1978) Distributive justice and utilitarianism. In: Arthur J, Shaw WH (eds) Justice and economic distribution. Prentice-Hall, New Jersey, p 103 et seq
Spangenberg J (2016–2017) Mehr wert als der Preis – Werte, externe Kosten und warum ihre Berechnung in Geldeinheiten mehr schadet als nutzt. Jahrbuch Nachhaltige Ökonomie:309 et seq
Spash C (2007) Deliberative monetary valuation. Ecol Econ:690 et seq
Stelzer T (2009) Corporate social responsibility. Berlin
Stern N (2009) A blueprint for a safer planet. How to manage climate change and create a new era of progress and prosperity. Bodley Head, Cambridge
Suchanek A (2000) Normative Umweltökonomik. Tübingen, Mohr Siebeck

Susnjar D (2010) Proportionality, fundamental rights, and balance of powers. Martinus Nijhoff Publishers, Leiden

TEEB (2009) The economics of ecosystems and biodiversity: TEEB for national and international policymakers

TEEB (2010) The economics of ecosystems and biodiversity: the ecological and economic foundations

Umweltbundesamt (UBA) (1997) Nachhaltiges Deutschland. Berlin

Unnerstall H (1999) Rechte zukünftiger Generationen. Königshausen und Neumann, Würzburg

von Egan-Krieger T (2014) Die Illusion wertfreier Ökonomie. Eine Untersuchung der Normativität heterodoxer Theorien. Campus Verlag, Frankfurt a.M. et al

von Kutschera F (1982) Grundlagen der Ethik. De Gruyter, Berlin

Wieland J, Schmiedeknecht M, Heck A (2009) Komparative Analyse: Informal und Regional – corporate social responsibility bei kleinen und mittleren Unternehmen. In: Wieland J (ed) CSR als Netzwerkgovernance – theoretische Herausforderungen und praktische Antworten. Über das Netzwerk von Wirtschaft, Politik und Zivilgesellschaft. Metropolis-Verl, Marburg, p 67 et seq.

Wilhelmi R (2012) Konsens durch Effizienz – Ordnungsökonomik als Utilitarismus? In: Aufderheide D, Dabrowski M (eds) Effizienz oder Glück? Wirtschaftsethische und moralökonomische Perspektiven der Kritik an ökonomischen Erfolgsfaktoren. Duncker & Humblot, Berlin, p 95 et seq

Wilson E (2012) The social conquest of earth. Liveright Publishing Corporation, New York

Zimpelmann B, Zöckler M (2008) Corporate social responsibility. Privates Regieren als Alternative für globale Sozial- und Umweltpolitik? In: Brunnengräber A, Burchardt, H-J, Görg C (eds) Mit mehr Ebenen zu mehr Gestaltung? Multi-Level-Governance in der transnationalen Sozial- und Umweltpolitik. Baden-Baden, p 119 et seq

6 Remaining Relevance of Cost-Benefit Elements in Balancing

Abstract

A clearly deflated cost-benefit analysis could (despite all limitations) contribute factual material (!) – in the form of actually monetarizable partial aspects of decision consequences – to ethical or legal balancing procedures. However, it cannot replace ethical and legal decision-making procedures or compete with them. The disciplinary peculiarities of economics, among other things, make it quite unlikely that economic cost-benefit analysts will be open to more modesty in this sense.

6.1 Sociology: Why Is Insight into the Limits of Common Economics so Little Received?

The idea that normativity can be framed in an empiricist way and that the advantages and disadvantages of measures can be determined based on preferences with the result of an empirically derived optimal state is, after all, untenable. The same is true with regard to the self-image cultivated by many economists that they do not practice ethics at all. Moreover, the basic principles of ethics, although normative, turned out to be quite objectively specifiable and not necessarily dependent on factual preferences. It is not possible to calculate the ethically and legally correct degree of sustainability, such as climate protection, including all associated follow-up questions. Rather, one can (also) specify leeway in climate policy with the derivable balancing rules, in which every decision can be considered acceptable. Within this balancing leeway, one simply has to make decisions, worldwide and also nationally. Furthermore, the concrete decision of individual ethical questions shows too much fuzziness to identify exclusively one possible answer to individual questions in concrete life as rational; however, the balancing rules and the institutional

F. Ekardt, *Economic Evaluation, Cost-Benefit Analysis, Economic Ethics*, Environmental Humanities: Transformation, Governance, Ethics, Law 2,
https://doi.org/10.1007/978-3-030-99284-2_6

responsibilities, which more closely contain the scope for decision-making in the case of precisely those fuzziness, can again be specified objectively. Rules for everyday individual situations, however, can only emerge from this in such a way that a (parliamentary) legislator enacts a (very) large number of laws in filling the leeway, as is characteristic of every modern legal system. The human capacity for rationality is empirically overestimated in the neoclassical mainstream. Economists in particular have pointed this out repeatedly in recent decades, and sociologists and psychologists even more so.[1] By the same token, normative rationality in its principle existence is underestimated by economists and at the same time overestimated in its potential to deliver concrete decisions for individual cases.

At the same time, caution is called for when someone brings the quite detailed critique of economization in the form of cost-benefit analysis under large-scale concepts such as alienation or colonization of the lifeworld, as this may seem sensible to some in the wake of Jürgen Habermas and Theodor Adorno (or, in the last instance, Karl Marx or even Jean-Jacques Rousseau).[2] The problems presented in the course of the analysis are too specific for that. Moreover, under broad headings such as alienation or colonization, other things are often criticized that are perhaps not worthy of criticism at all – as was discussed above for the economic policy instruments. Moreover, especially in the older critical theory of Adorno, there is no clear and tenable justification of the presupposed critical standards, as Habermas has already worked out in detail.[3]

However, any criticism of cost-benefit analysis (unless it merely promises to improve its methods and thus its application in detail) has so far largely bounced off neoclassical economics. In this respect, economics, like jurisprudence, has proved to be a discipline with particularly strong orthodox tendencies.[4] This raises the question – beyond the discussion about reasons for or against cost-benefit analysis and beyond the consideration of its historical genesis – how this camp thinking in economics comes about. Primarily, a separated discourse exists in modern science in general between the disciplines and especially with regard to environmental protection.[5] Thus, especially objections that are seen as non-economic have a hard time. The problem that in such talk of disciplinary truths a relativism shines through that is not convincing has already been discussed (Sects. 1.4 and 5.1). Notwithstanding this, a high degree of pure self-reference persists in the relevant economic literature, with the result that ethical and legal analyses are hardly noticed. Accordingly, even in a very careful and nuanced work on the economic perspective, one finds the startling statement: “There is little literature to date that discusses ethical aspects of

[1] Simon 1993; Steinberg 2013, p. 102 et seq.; Rogall 2013, p. 110 et seq. and 213.

[2] See on this Horkheimer and Adorno 2012; Adorno 2015; Habermas 1981, Vol. 1–2.

[3] Habermas 1981, Vol. 1, 1983, p. 56 et seq. Furthermore, in the older Critical Theory there is a far reaching over into the realm of the good life, up to quasi eschatological-messianic ideas of an end of time and a preaching of more love. Another problem is that irrational methodologies like the dialectically meant rigid negation (Bestimmte Negation) are practiced there.

[4] Dürmeier et al. 2006; Hühn 2015, p. 101 et seq.; Schlaudt 2016, p. 18 et seq.

[5] See also kritisch Gawel 2000, p. 10.

risk."[6] The entire debate on environmental ethics and environmental law concerning such balancing problems is thus apparently largely unknown.[7]

How such quite peculiar views, states and processes stabilize is again a question for the behavioral sciences. In this respect, it is always surprising that economists sometimes seem to apply little of their own findings, such as those of behavioral economics research – which is, after all, part of the behavioral sciences – to themselves. Emotional factors such as group dynamics, the stability of a certain socialization, the feared loss of face in the case of admitting mistakes (as well as the threat of damage to careers) or the adherence to established conceptions of normality are at play here, as elsewhere in human behavior. The stability of a point of view and its comparison in a closed network thus offers no real guarantee that what is commonly assumed in this network is actually true.[8] Admittedly, the empiricist paradigm fits in to a certain extent with other contemporary tendencies, such as the dominance of the natural sciences or digitization, and, as indicated, it is in no way limited to the economic sciences, but dominates today's understanding of science as a whole. In this respect, it seems not unlikely that criticism of the cost-benefit analysis that goes beyond cautious and not too dismissive admissions on questions of application will continue to be dismissed in the future.

6.2 Why a Deflated Cost-Benefit Analysis Nevertheless Remains Useful for Ethics and Law – And a Conclusion

Criticism of cost-benefit analysis, as already indicated, is not tantamount to a call to drop cost-benefit analysis altogether (and anyway, the ethics of cost-benefit analysis in particular should be thanked for constantly pointing out the need for balancing in all decisions[9]). In conclusion, we will now briefly discuss one such deflated version of cost-benefit analysis, which at the same time remains very relevant.[10] One has to give up the intention of a comprehensive approach for this (this comprehensive notion is also contained in the term external costs, popular among environmentalists, which suggests that one can fully translate non-monetary environmental effects into monetary values; it should therefore be abandoned).[11]

[6] See Fehling 2009, p. 188.

[7] Also the concrete explanations in Fehling 2009, p. 144 et seq. and Jones-Lee 1976 do not really help, since they do not address the problems discussed above in Chaps. 2, 3, 4 and 5. Also, the quantification problems are not really discussed in a meaningful way, since in some cases it is only argued as if life is to be compared with life; however, one of the problems is precisely to compare different goods and whether this can be done by monetarizing or not.

[8] Unclear on this Enders 2008; explicitly for a transdisciplinary (respectively determined by content-related questions and not by disciplinary boundaries) search for knowledge Ekardt 2019, Ch. 1.7 (furthermore there in Chap. 2 on the respective behavioral science findings).

[9] See Fehling 2004.

[10] In this direction also Winter 2001, p. 124; Fehling 2004; similarly, Meßerschmidt 2001; rather vague hints can be found in Britz 1997 and Wittmann 2004, p. 143 et seq.

[11] Aptly on this Spangenberg 2017. In contrast, it is presupposed in Hey 2012, p. 125 (128 et seq.).

Despite the many criticisms, ethical and legal balancing (Sect. 4.3) and economic cost-benefit analysis are not purely exclusive. Rather, there are elements of complementarity, even if cost-benefit analysis can neither compete with the broad ethical and legal balancing framework nor replace it in its binding nature:

- The entire criticism formulated above does not exclude, within legal or ethical balancing rules, to name real existing market prices of a promotion or damage of certain legal goods. Rather, this is even helpful and absolutely to be welcomed to better grasp the degree of impairment of an interest. Thus, if it can be said which damages to buildings cause twice as frequent floods, then the public authorities must receive this as such – for the sake of the balancing rule, which requires careful fact-finding.[12] This is then an important information for determining the weight of, for example, freedom of property in the necessary balancing. The goal of fact-finding and participation (via real prices paid on the market), as commonly attributed to cost-benefit analysis, can thus be taken up.[13]
- Furthermore, within the objective framework of the ethically and legally founded balancing rules, the national or transnational legislator may also make his leeway for subjective weightings more transparent and comprehensible by assigning a numerical value to normative concerns; but this is then a subjective decision and has nothing objective about it, and this subjective decision is then not to be presented by an economist as supposedly objective, but to be made by a democratically legitimized politician.[14] Such voluntary quantification of the weight of concerns may not be strictly required and often not easily feasible, but it is sometimes helpful; it may also make it easier for the courts to verify compliance with some of the rules of balancing.
- At the same time, this means that governance research will continue to be useful for calculating the cost-related efficiency (see Sect. 1.2) of instruments measured against defined targets and limited to target-relevant effects that have a market price. At the same time, this can have a positive effect on the actual acceptance of more far-reaching ecological measures, because in this way it can become transparent that climate change, for example, threatens to become far more expensive than effective climate protection would be.[15] As we have seen, economic research contributions to behavioral research or to economic policy instruments remain quite untouched by the criticism of cost-benefit analysis (Sect. 2.2.8).

---

[12] For example, on the basis of flood protection Hirschfeld 2012, p. 277 et seq.; on the discussion of private pre-financing of roads Ekardt 1997; on the discussion of usage costs for water bodies Gawel 2014.

[13] See Pearce et al. 2006, p. 34 et seq.; Nestle 2012, p. 104 et seq.; Hansjürgens and Lienhoop 2015, p. 18 et seq.; Hanley and Barbier 2009, p. 15 et seq.

[14] See Hofmann 2007, p. 251 et seq. This voluntary quantification may have advantages in terms of equal treatment and verifiability.

[15] Classically again Stern 2007; on the various forms of cost accounting Ekardt 2017a, b, p. 9 et seq.

- For the private sector, most of the problems of cost-benefit analysis developed here do not arise in this way. For as long as they remain within the framework of the law, companies or individuals are free to make their own abstract or concrete decisions. The point that this strategy may not be consistent with basic democratic requirements for public authority or that it has major application problems need not bother a firm (or an individual), unlike the state. And fundamental theoretical concerns about the basic ethical positions behind the cost-benefit analysis do not arise on a purely individual or entrepreneurial level. This is true at least if one assumes, as the author does, that an individual ethics (or even an ethic of the good life) cannot be formulated in any case.[16]

At these points, the relevant research is and remains undoubtedly very valuable and can lead to unvarnished, realistic observations of facts. However, neoclassical economics should take its own intentions more seriously than it has done so far. If one wants to determine numerical values, one should at any rate try to state completely the really monetary costs and benefits, as far as they can be well grasped in market prices. This would then show that concrete monetary climate damage such as crop failures or storm damage (and even more so climate wars) would be more expensive than an effective climate policy.[17] Here lie central achievements of the IPCC reports and, e.g., also of the Stern Report. Statements about the probability of occurrence of events, on the other hand, can often only be provided by economists and natural scientists with a lower degree of precision than one might hope for; the natural conditions of climate change and the world economy are simply too complex for this. Thus, of course, the factors just mentioned are more suitable as soft information, but not as a determining factor.

A perhaps more modest, no longer normative, moreover less extensively quantifying and less natural-science-oriented, in short: an economic sustainability research more transdisciplinarily merged with the other sustainability human sciences could be a worthwhile goal in all this.[18] The empiricist creed that only facts – quantifiable and reproducible ones – should be taken into account must be reconsidered. Figures really do not make complex balancing unnecessary. This is also true, for example, when it comes to measuring prosperity differently in the future.[19] The analogous points of criticism as for efficiency approaches also apply, as seen, to popular attempts among natural scientists to always express normative goals, for example,

[16] See in detail with further references Ekardt 2021, §§ 3 A., 4 F. IV., 6 B.; Ekardt 2019, Ch. 3.1 and 4.2.

[17] Stern 2007, p. 151 speaks only in general terms of possibly increasing instability. Binswanger 2010, p. 154 et seq. is very critical of the fact that a massive effort to quantify and formalize is coupled with sometimes very obvious omissions in economic approaches.

[18] It is therefore to be welcomed when Stern, as the most publicly perceived climate economist, himself at least partially acknowledges the omissions of the economic approach (even if only in a sweeping way and without addressing the basic problems of growth and preference theory); Stern 2007, p. 149 et seq.

[19] Critical of the often underlying ethical capability approach (on this, e.g., Sen 2009) Ekardt 2019, Ch. 3.2.

in terms of biodiversity or ecosystem states, in numerical form and to justify them normatively based on empirical facts, especially since there are further reservations against such sustainability indicators and the like. The difficulty of making such statements, however, does not mean that objective normative statements are not possible in life. Sustainability and other societal goals thus remain more than arbitrary subjective settings. At the same time, all this has shown that liberal democracy and all of us could benefit in the longer term if some aspects of the rampant number crunching were to be given a question mark.

Overall, this volume leads to the following conclusion. The objections raised in the tradition criticism of utilitarianism against the economization of decision-making, such as the critique of the balancing orientation, of allegedly lacking distributive justice, or of the tendency to describe human beings as self-interested in terms of behavioral science, are not convincing upon closer examination. Beyond that, however, cost-benefit analysis in its present form faces insoluble problems of application. Furthermore, it has no viable basic theory and is difficult to reconcile with liberal democratic constitutional law. All of this applies both to overall societal considerations and to individual analyses, such as the construction of a coal-fired power plant. A clearly deflated cost-benefit analysis could nevertheless contribute factual material to ethical or legal decision-making procedures, namely in the form of partial aspects of decision consequences that can actually be depicted in monetary terms.

Repetition Questions

1. What problems typically arise in academic disciplines when they are confronted with fundamental critique? (Sect. 6.1).
2. Can the criticism of cost-benefit analysis be equated with a critique of a general economization of society? (Sects. 6.1 and 2.2.8).
3. In what way could cost-benefit analysis nevertheless play a very important role despite all the criticism? (Sect. 6.2).

References[20]

Adorno T (2015) Minima Moralia, new edition. Darmstadt

Binswanger M (2010) Sinnlose Wettbewerbe. Freiburg

Britz G (1997) Umweltrecht im Spannungsverhältnis von ökonomischer Effizienz und Verfassungsrecht, Die Verwaltung:185 et seq.

[20] In accordance with legal practice, parliamentary, governmental and EU Commission documents as well as laws and judgments are not listed in the bibliography, as they can be found unequivocally based on the reference given in the continuous text or via the general search engines. The last access date for all Internet sources is 31.10.2021.

Dürmeier T, von Egan-Krieger T, Peukert H (eds) (2006) Die Scheuklappen der Wirtschaftswissenschaft. Postautistische Ökonomik für eine pluralistische Wirtschaftslehre. Metropolis-Verl, Marburg

Ekardt F (1997) Zur Verfassungsmäßigkeit der sogenannten Privatfinanzierung von Verkehrswegen. Verwaltungsblätter für Baden-Württemberg:281 et seq

Ekardt F (2017a) Natural capital accounting and the mobilization of financial resources for biodiversity, Analyses for the German Federal Government. Leipzig. http://felix-ekardt.eu/files/texts/Biodiv-Evaluation-Instruments.pdf

Ekardt F (2017b) Wir können uns ändern. Gesellschaftlicher Wandel jenseits von Kapitalismuskritik und Revolution. München

Ekardt F (2019) Sustainability. Transformation, governance, ethics, law. Springer, Dordrecht

Ekardt F (2021) Theorie der Nachhaltigkeit. Ethische, rechtliche, politische und transformative Zugänge – am Beispiel von Klimawandel, Ressourcenknappheit und Welthandel, 4th edn. (= 3rd ed. of the new edition). Baden-Baden

Enders J (2008) Wissensnetzwerke in der Klimapolitik. Frankfurt a.M

Fehling M (2004) Kosten-Nutzen-Analysen als Maßstab für Verwaltungsentscheidungen. Verwaltungsarchiv:443 et seq

Fehling J (2009) Die Ethik des Value of a Statistical Life. Die Rolle individueller Risikokompetenz für die Legitimität des VSL. Rainer Hampp Verlag, Munich

Gawel E (2000) Effizientes Umweltordnungsrecht – Anforderungen und Grenzen aus ökonomischer Sicht. In: Gawel E, Lübbe-Wolff G (eds) Effizientes Umweltordnungsrecht. Kriterien und Grenzen. Nomos, Baden-Baden, p 9 et seq

Gawel E (2014) Konzept zur „Berücksichtigung" von Umwelt- und Ressourcenkosten nach Art. 9 WRRL, Verwaltungsarchiv:99 et seq

Habermas J (1981) Theorie des kommunikativen Handelns, 2 Bde. Frankfurt a.M

Habermas J (1983) Moralbewusstsein und kommunikatives Handeln. Frankfurt a.M

Hanley N, Barbier E (2009) Pricing nature. Cost-benefit analysis and environmental policy. Elgar, Cheltenham

Hansjürgens B, Lienhoop N (2015) Was uns die Natur wert ist. Potenziale ökonomischer Bewertung. Metropolis-Verlag, Marburg

Hey C (2012) Wege aus dem Wachstumsdilemma. Kritische Anmerkungen zu einer aktuellen Debatte. Zeitschrift für Umweltpolitik und Umweltrecht 35:125 et seq

Hirschfeld J (2012) Kosten und Nutzen der Anpassung an den Klimawandel – ein Verteilungsproblem. In: Ekardt F (ed) Klimagerechtigkeit. Ethische, rechtliche, ökonomische und transdisziplinäre Zugänge. Metropolis-Verlag, Marburg, p 277 et seq

Hofmann E (2007) Abwägung im Recht. Chancen und Grenzen numerischer Verfahren im öffentlichen Recht. Mohr Siebeck, Tübingen

Horkheimer M, Adorno T (2012) Dialektik der Aufklärung, New edition. Frankfurt a.M

Hühn M (2015) Wie die Wirtschaftswissenschaften sich der Ethik entledigten. Jahrbuch Ökonomie und Gesellschaft:101 et seq

Jones-Lee M (1976) The value of life. An economic analysis. University of Chicago Press, London

Meßerschmidt K (2001) Ökonomische Effizienz und juristische Verhältnismäßigkeit – Gemeinsames und Trennendes. In: Gawel E (ed) Effizienz im Umweltrecht. Grundsatzfragen einer wirtschaftlichen Umweltnutzung aus rechts-, wirtschafts- und sozialwissenschaftlicher Sicht. Nomos, Baden-Baden, p 215 et seq

Nestle I (2012) The costs of climate change in the agricultural sector. A comparison of two calculation approaches, Dissertation, Flensburg

Pearce D, Atkinson G, Mourato S (2006) Cost-benefit analysis and the environment. Recent development. Paris

Rogall H (2013) Volkswirtschaftslehre für Sozialwissenschaftler. Einführung in eine zukunftsfähige Wirtschaftslehre, 2nd edn. Springer, Wiesbaden

Schlaudt O (2016) Wirtschaft im Kontext. Eine Einführung in die Philosophie der Wirtschaftswissenschaften in Zeiten des Umbruchs. Vittorio Klostermann GmbH, Frankfurt a.M

Sen A (2009) The idea of justice. Harvard University Press

Simon H (1993) Homo rationalis. Frankfurt a.M

Spangenberg J (2016–2017) Mehr wert als der Preis – Werte, externe Kosten und warum ihre Berechnung in Geldeinheiten mehr schadet als nutzt. Jahrbuch Nachhaltige Ökonomie:309 et seq

Steinberg R (2013) Die Repräsentation des Volkes. Menschenbild und demokratisches Regierungssystem. Nomos, Baden-Baden

Stern N (2007) Stern review. The economics of climate change. HM Treasury, Cambridge

Winter G (2001) Über Nutzen und Kosten der Effizienzregel im öffentlichen Recht. In: Gawel E (ed) Effizienz im Umweltrecht. Grundsatzfragen einer wirtschaftlichen Umweltnutzung aus rechts-, wirtschafts- und sozialwissenschaftlicher Sicht. Nomos-Verl.-Ges, Baden-Baden, p 97 et seq

Wittmann BK (2004) Effizienzsteigerung bei Gerichten im Spannungsfeld zwischen richterlicher Unabhängigkeit und Anwendung betriebswirtschaftlicher Methoden. Munich

Summary

The economic evaluation practiced in economics and beyond is an attempt to make correct individual or collective decisions derivable in a simple way, despite the immense complexity of many decision situations. Its primary form, the cost-benefit analysis, has been the topic of this volume with particular regard to its chances and limits. Cost-benefit analysis is an attempt to raise the time-honored philosophical, legal and political problem of right human decision-making to a new level and to make it more precise, clearer and thus more rationally treatable. It is thus ultimately an attempt to inherit practical philosophy or ethics and to bring it into a supposedly more objective form. Cost-benefit analysis refers to the attempt to identify optimal/efficient/desirable social conditions based on people's preferences that are factually available and, if possible, translated into monetary values. This method, which is also increasingly used in politics (especially in environmental policy), takes people's factual preferences as they are. Based on this, it tries to identify social conditions with the greatest possible benefit which are translated into monetary units and the lowest possible cost. This can be understood as a sign of the highest possible individual satisfaction of needs.

To identify factual preferences and to assign a monetary value to them, cost-benefit analysis determines the willingness to pay for a certain condition, for example, for more or less climate protection. With regard to the exact determination of preferences, cost-benefit analysis acts as umbrella over a variety of methodological approaches that can be applied individually or in combination. These assess, often elicit, preferences directly (from a limited group of individuals), and hypothetically, for example in terms of willingness to pay for a possibly longer (or less short) life due to increased climate action. Furthermore, it is observed whether certain advantages and disadvantages (respectively benefits and costs) of possible decisions on the market already have a market price or whether such a price can be implicitly derived from the actions of individuals.

According to their intention, cost-benefit analyses of social welfare can be applied in different places. Cost-benefit analyses are relevant for abstract-general decisions and for individual decisions, and they are relevant for the public sector as well as for the private sector. Thus, one can measure an act of parliament or a

F. Ekardt, *Economic Evaluation, Cost-Benefit Analysis, Economic Ethics*, Environmental Humanities: Transformation, Governance, Ethics, Law 2, https://doi.org/10.1007/978-3-030-99284-2

general business strategy against them as well as an individual government action or an individual business decision. In this volume, abstract-general decisions in the public sector are considered, exemplified by climate policy (as a central example of sustainability policy, i.e., a policy that aims at intertemporally and globally tenable ways of living and doing business). It does not extensively consider other forms of economic evaluation beside cost-benefit analysis such as national accounts (which do not comprehensively present the advantages and disadvantages of decisions in monetary terms, but rather list them in an additive manner) or cost efficiency analysis (which primarily examines real costs in the sense of effects in market prices in the pursuit of a politically determined goal, i.e., does not comprehensively focus all advantages and disadvantages of various possible decisions in monetized form, as does cost-benefit analysis). In contrast, this volume briefly considers so-called economic ethics, which operates without monetization, but nevertheless with a derivation of decisions based on factual preferences of individuals and finds some appeal especially among economists, but at times also in other disciplines. In a broader sense, the work thus fits into the discourse on economization, financialization, monetarization, commodification, commercialization, valorization, and fiscalization as phenomena of the modern world, although these vague and ambiguous concepts deliberately do not have a stronger focus here.

Methodologically, this thesis approaches the topic on several levels. After various epistemological preliminaries and a literature analysis on the historical genesis of cost-benefit analysis and its precursors in the history of philosophy, the thesis asks whether an existing methodology (namely that of cost-benefit analysis as an implicitly practical philosophical theory) is feasible at all by going through the application of that method. Then its legal conformity is analyzed. This requires the use of the common methods of legal interpretation, (i.e., the rules of interpretation of literal sense, systematics, norm purpose and history of norm formation). Finally, the normative background assumptions of cost-benefit analysis are analytically identified and critically examined as a covert practical-philosophical theory or ethics. Essentially, ethics methodically deals with the correct argumentation, taking into account especially the basics of epistemology and logic, of course also in discussion with arguments already developed by other authors, etc. Because of their special importance, certain epistemological distinctions are explained and justified in advance, in particular the distinctions between objective/subjective and being/should (contrary to common opinion), which are not parallel, but rather transverse to each other. Thus, for facts as well as for norms it can be asked whether they can be recognized objectively. Despite certain limitations, both can ultimately be affirmed. In addition to the epistemological analyses and the application analyses, the examination of legal conformity and the philosophical examination of the fundamentals, the work also considers certain widespread, but precisely unconvincing points of criticism of cost-benefit analysis in detail.

Although economists do not experience their approach as normative, on closer examination it can be identified as an ethics based on economic science. Theoretically, it stands in the tradition of empiricism, which limits the realm of the knowable to facts (rather than of knowable norms as well), and if possible, to reproducible and quantifiable facts. In this respect, lines in the history of philosophy can be drawn to cost-benefit analysis in twentieth century (and its economic foundation, so-called neoclassicism) already from medieval nominalism via the Reformation and Scottish moral philosophy. It would be too short-sighted to associate cost-benefit analysis solely with philosophical utilitarianism – with which there are some differences.

The objections raised against cost-benefit analysis in the tradition of the criticism of utilitarianism, such as the criticism of its orientation towards balancing, of an allegedly lack of distributive justice, or of the tendency to describe human beings as egoist in terms of behavioral science, are hardly convincing upon closer examination. In short, for various reasons, balancing is unavoidable in decision-making; absolutely balancing-free normative statements cannot be identified. Even the overestimated distinction between deontology and consequentialism (which is not feasible) does not change this. With respect to the behavioral description of human beings, on the other hand, economic research is more complex than sometimes assumed; and in general, this kind of description has also nothing directly to do with the (normative) cost-benefit analysis. Other objections against the cost-benefit analysis, which are not resounding on closer examination, can be named with the keywords quantification, informedness, concept of utility, individualism, and criticism of capitalism or modernity.

In the debate as a whole, cost-benefit analysis and policy goal-setting and balancing are linked in parts with the debate on the most effective instruments, e.g., of climate policy. For example, in the interest of climate protection, levies or environmental certificate markets can be conceived as instruments to internalize external costs. In the environmental context, economic policy instruments ultimately aim to induce people to do more to protect the environment by means of price signals. However, no one says that the desired limitation of environmental consumption must target a quantity that must first be determined by a cost-benefit analysis. In this respect, cost-benefit analysis and economic instruments are two separate objects.

Although the above-mentioned well-known arguments against cost-benefit analysis are unconvincing, cost-benefit analysis fails because of certain unsolvable application problems. The frequently debated issues of dealing with risk, uncertainty and discounting of future concerns are by no means the only relevant aspects, but the very uncertainty of many cost-and-benefit effects mathematically precludes a monetarily exact quantification of decision situations. Even if leaving this aside, economists do not have a reliable way to represent the various cost-and-benefit effects that are not available in the form of market prices in a truly monetary way, despite numerous attempts. Inquiring about a hypothetical willingness to pay or supposedly observing the behavior of individuals (morals of the markets) leads to very incomplete, fictitious and, among other things, dispersed findings also due to very differently distributed abilities to pay. Empirically, the belief in infinite economic growth underlying cost-benefit analysis appears rather less robust. Thus, there are unsolvable application problems for the cost-benefit analysis. Those problems especially (but not exclusively) become visible in the context of climate protection, both in overall considerations of societal problems and in individual analyses, such as the construction of a coal-fired power plant.

In addition, there are collisions between cost-benefit analysis and liberal-democratic constitutional law, the principles of which are also fundamental ethical principles. This concerns both the rights to freedom and its elementary preconditions as well as the model of democracy and respect for the rule of law. The weight of the fundamental rights and the balancing rules applicable in the case of conflicting spheres of freedom in liberal democracies – which, in short, form the basis of the legal decision-making scheme (then further concretized in law in many details) – do not depend on how much someone can pay. Freedom in a liberal democracy is not only the freedom of solvent consumers with their preferences, but also the freedom of those who are hardly present in the free market due to a lack of purchasing power and who would suffer massively from, for example, the consequences of climate change and damaged ecosystems. Furthermore, for good reasons, decision-making in liberal democracies is representative-democratic with only isolated plebiscitary (and participatory) additions. This is contrary to the plebiscitary intentions of the cost-benefit analysis. Also, for good reasons, confirmed by centuries of historical experience, modern law claims a monopoly of the decision of normative questions for itself, which only knows an opening in favor of other decision-making procedures, such as the free play of forces, where this is permitted by law. A competing normativity to the law on the part of the cost-benefit analysis is thus excluded.

Furthermore, the empiricist theoretical basis of (covertly practical-philosophical, normative) cost-benefit analysis is untenable. This theoretical basis consists of the empiricism-oriented idea that normative questions have to be transformed into questions of factual preferences. In this context, there are logically false conclusions and performative contradictions, among other things because the idea that normative questions per se are not objectively decidable leads to self-contradictions. Furthermore, cost-benefit analysis does not hold up to any application to itself and, upon closer examination, can hardly make plausible how it actually fits its own empiricist foundations. Cost-benefit analysis also fails to find a convincing approach to intertemporal justice. Moreover, the economist attempt to deflate ethics into a question of factual preferences for altruism is unsuccessful. Similar problems also arise for so-called economic ethics, which, apart from monetization, just runs largely parallel to cost-benefit analysis.

A clearly deflated cost-benefit analysis could nevertheless contribute factual material (!) – in the form of actually monetizable partial aspects of decision consequences – to ethical or legal balancing procedures. However, it cannot replace ethical and legal decision-making procedures or compete with them. The disciplinary peculiarities of economics, among other things, make it quite unlikely that economic cost-benefit analysts will be open to more modesty in this sense.

Glossary

Balancing means weighing different normative goods. It is inevitable with regard to normative decisions, ethically and legally. However, the economic cost-benefit analysis is not a convincing balancing theory (Sects. 2.2.3, 2.2.4, and 4.3).

Balancing rules can be derived from the liberal basic principles of reason, dignity, impartiality, freedom, democracy and separation of powers. They frame the scope of objective statements possible regarding normativity (Sect. 4.3).

Capitalism does not consist solely of the principle of growth, nor is it identical with market economy. The criticism of capitalism partly makes false assumptions and overlooks its potentials – and at least cost-benefit analysis is not necessarily connected to capitalism (Sect. 2.2.9).

Cooperation can be driven by self-interest or altruism. Some overestimate and some underestimate the importance of cooperation for human behavior, inter alia with regard to sustainability (Sect. 2.2.1).

Culture and biology are the background to all motivational factors – the dispute over sociobiology shows that the dominant desire to latently deny one of the two factors is flawed (Sect. 2.2.1).

Consistency, (technological) efficiency and frugality represent three sustainability strategies. The first two focus on technically smarter production and consumption (consistency, for example, through recycling). Frugality, on the other hand, focuses on less production and consumption (Sect. 1.3).

Corporate social responsibility and sustainable consumption as an instrument of sustainability rely on voluntary action by companies and consumers. These approaches take too little account of the interaction between the actors, the motivational situation and the difficulties of concretization on the level of single actors (Sect. 5.2).

Economic policy instruments working by means of direct or indirect pricing of resources or sinks – can be designed as quantity control and thus adequately address sustainability as a quantity problem – and they also fit better with human motivation and the liberal basic principles than other governance instruments (Sect. 2.2.8).

F. Ekardt, *Economic Evaluation, Cost-Benefit Analysis, Economic Ethics*, Environmental Humanities: Transformation, Governance, Ethics, Law 2,
https://doi.org/10.1007/978-3-030-99284-2

Empiricism means an epistemology and at the same time a sceptic theory of justice, which are both not able to convince in essential respects (Sect. 1.1).

Environmental humanities means research in disciplines such as economics, law, philosophy, political science, sociology, cultural studies, ethnology, etc. on questions of environment and sustainability – preferably from a transdisciplinary perspective (Preface).

Epistemology means the theory of what can be objectively recognised in facts and, if necessary, in norms. This also includes basic distinctions such as objective versus subjective, is versus ought, genesis versus validity (which do not coincide) – and a critique of some aspects of constructivism (Sect. 1.4).

Fossil fuels (oil, coal, gas) are the main driver of various environmental problems such as climate change, loss of biodiversity, human diseases, disturbed nitrogen cycles, etc. due to their omnipresence in electricity, heat, mobility, plastics and in the agricultural sector (Sect. 1.3).

Freedom is the object of human rights. Under the auspices of sustainability, the understanding of freedom must be expanded, inter alia to include the elementary preconditions of freedom, an intertemporal and global dimension and an element of precaution (Sects. 4.2, 4.3, and 5.1).

Good life (roughly equivalent to happiness) is the antonym to justice and describes an area that is legally and ethically inaccessible to any regulation (Sect. 5.1).

Governance is about effective means – i.e., policy instruments – to implement policy objectives (and to balance them against other values or objectives) and sustainability strategies resulting from the objectives (Preface).

Governance problems such as rebound effects or shifting effects are typical obstacles to effective sustainability policy that can be empirically measured and derived from behavioral research (Sect. 2.2.8).

Happiness is empirically investigated in happiness research. The findings are ambivalent in terms of sustainability – and there is nothing normatively that can be said about happiness in liberal democracy (Sect. 2.2.1).

Human rights mean (ethically and in national, European and international law) the rights to freedom and the elementary preconditions of freedom. Human rights and fundamental rights are synonymous for the purposes of this book (Sects. 4.2, 4.3, and 5.1).

Individual versus structure (or micro versus macro) is a popular distinction in research on behavior and societies, but it is ultimately not feasible.

Interconnectedness refers to the way politicians, citizens, consumers, entrepreneurs and other actors influence and depend on each other. It is thus a chicken-and-egg game to ask for the main responsible entity (Sect. 2.2.1).

Integrated solutions for various sustainability challenges such as climate change, biodiversity loss or disturbed nitrogen cycles are necessary. What is needed is addressing the overarching drivers such as fossil fuels and livestock farming (Sects. 1.1, 1.3, and 2.2.8).

Justice means the rightness of social order and human behavior. In contrast, truth means the accuracy of statements about facts. Social distributive justice is a sub-aspect of justice regarding distributional issues (Sects. 1.4 and 2.2.6).

Knowledge and values are constantly overestimated as motivational factors of human behavior. But if one does not ask descriptively what drives us in fact, but asks normatively what we should do, values are the sole yardstick (Sect. 2.2.1).

Law and ethics are normative systems, of which law is characterized by greater concreteness and the existence of state sanctions (Sect. 1.4).

Liberal democracy is based on the liberal basic principles of reason, dignity, impartiality, freedom, (representative) democracy and separation of powers (Sects. 4.2, 4.3, and 5.1).

Liberalism, skepticism, contextualism, and metaphysics are different basic approaches to the theory of justice. Only reason-based liberalism can convince (Sect. 2.2.2).

Livestock farming is – besides fossil fuels – the second major driver of various environmental problems (Sects. 1.2 and 4.9).

Marxism is a theory which tries to overcome the distinction between is and ought, but without success – and which empirically and normatively encounters many frictions (Sects. 2.6 and 3.1).

Methodology with regard to sustainability must be multi-methodical and qualitatively oriented in essential parts, for example, when it comes to human behavior and governance. The methodology is different when it comes to normative questions, e.g., the interpretation of the law (Sects. 1.4 and 2.2.1).

Motivational factors help explain the behavior of citizens, consumers, politicians, entrepreneurs and other actors. They include, e.g., self-interest, values, emotions, conceptions of normality, knowledge and structural conditions such as path dependencies and problems of collective goods (Sect. 2.2.1).

Paris Agreement is the most important global environmental agreement, which is vague in many respects, but sets a very ambitious – and legally binding – climate target in its Article 2 (Sects. 1.3 and 4.3).

Ping-pong of change means that social change is possible in an interplay of interdependent social actors such as politicians, citizens, consumers, managers, etc. (Sect. 2.2.1).

Post-growth or degrowth means an economy without economic growth or even with processes of shrinking – possibly necessary for sustainability, but with considerable and partly unsolved consequences (Sect. 3.4).

Rationality or reason means the ability to decide questions based on reasons. A distinction must be made between theoretical, instrumental and normative reason (Sects. 1.4 and 2.2.2).

Sustainability means the expansion of ethics, law and politics in intertemporal and global terms, i.e., intertemporal and global justice. In contrast, a three-pillar model of sustainability is not convincing (Sect. 1.3).

Sustainability governance seeks an effective implementation of (human-rights-based) sustainability goals. Caps for fossil fuels and livestock farming play a key role for that (Sect. 2.2.8).

Sustainable Development Goals have been established by the UN for a bouquet of relevant policy areas. They are legally non-binding and partly contradictory.

Transdisciplinarity means discussing science not from the perspective of a particular discipline (and its unquestioned dogmas), but based on substantial problems – and thereby incorporating all disciplinary knowledge on these problems. Furthermore, transdisciplinarity is open to knowledge stocks outside the institutionalized scientific system, which is self-evident in an objective concept of rationality (Preface).

Uncertainty and risk are characteristic of sustainability issues – with regard to both facts and norms. There are ethical and legal rules for dealing with them (Sect. 5.3).

Universalism refers to justice in all societies – and globalism designates justice across borders between all societies (Sects. 2.2.2 and 5.1).

Bibliography

Alier J (2009) Socially sustainable economic de-growth. Dev Chang 40:1099 et seq
Altvater E, Brunnengräber A (2011) With the market against climate catastrophe – can that succeed? In: Altvater E, Brunnengräber A (eds) After cancun. Climate governance or climate conflicts. VS Verlag für Sozialwissenschaften, Wiesbaden, p 11 et seq
Anderson M, Teisl M, Noblet C, Klein S (2015a) The incompatibility of benefit–cost analysis with sustainability science. Sustain Sci 10(1):33–41. https://doi.org/10.1007/s11625-014-0266-4
Anderson M, Teisl M, Noblet C (2015b) Whose values count: is a theory of social choice for sustainability science possible? Sustain Sci 11(3):373–383. https://doi.org/10.1007/s11625-015-0345-1
Antes R (2014) Nachhaltigkeit und Betriebswirtschaftslehre: Eine wissenschafts- und institutionentheoretische Perspektive. Marburg
Aquinas T (1934) Summa Theologiae, new edition. Munich
Baer S (2006) „Der Bürger" im Verwaltungsrecht. Subjektkonstruktionen durch Leitbilder vom Staat. Tübingen
Ballestrem KG (2001) Adam Smith. Munich
Beck U (1974) Objektivität und Normativität. Die Theorie-Praxis-Debatte in der modernen deutschen und amerikanischen Soziologie. b.H. Rowohlt, Reinbek
Benz U, Fetzer S (2004) Was sind gute Nachhaltigkeitsindikatoren? OECD-Methode und Generationenbilanzierung im empirischen Vergleich. Inst. für Finanzwiss. der Albert-Ludwigs-Univ, Freiburg
Binswanger M (2001) Technological progress and sustainable development: what about the rebound effect? Ecol Econ 36:119 et seq
Binswanger M (2010) Sinnlose Wettbewerbe. Freiburg
Birnstein U (2009) Der Reformator. Wie Johannes Calvin Zucht und Freiheit lehrte, 2nd edn. Wichern-Verlag, Berlin
Bitter M (2011) Contradictions of the commodity carbon – on the material and symbolic production of a market. In: Altvater E, Brunnengräber A (eds) After cancun. Climate governance or climate conflicts. VS Verlag für Sozialwissenschaften, Wiesbaden, p 71 et seq
Bofinger P (2007) Grundzüge der Volkswirtschaftslehre, 2nd edn. Munich
Bröckling U (2007) Das unternehmerische Selbst. Soziologie einer Subjektivierungsform. Suhrkamp, Frankfurt a.M
Brookes L (1978) Energy policy, the energy price fallacy and the role of nuclear energy in the UK. Energy Policy 6:94 et seq
Brookes L (1979) A low energy strategy for the UK. Atom:73 et seq
Brugger W, Neumann U, Kirste S (eds) (2008) Rechtsphilosophie im 21. Jahrhundert. Suhrkamp, Frankfurt a.M

F. Ekardt, *Economic Evaluation, Cost-Benefit Analysis, Economic Ethics*, Environmental Humanities: Transformation, Governance, Ethics, Law 2, https://doi.org/10.1007/978-3-030-99284-2

Bundeswehr (2010) Peak Oil, Zentrum für Transformation der Bundeswehr – Dezernat Zukunftsanalyse. Strausberg

Carr N (2010) Wer bin ich, wenn ich online bin – und was macht mein Gehirn solange? Wie das Internet unser Denken verändert, 3rd edn. Random House, Munich

Carstensen K, Schefold B (1994) Die Klassische Politische Ökonomie. In: Issing O (ed) Geschichte der Nationalökonomie, 3rd edn. Munich, p 63 et seq

Chen S, Ravallion M (2008) The developing world is poorer than we thought, but no less successful in the fight against poverty, World Bank development research group, Working paper No. 4703. World Bank, Washington, DC

Clarke B, Foster J (2001) William Stanley Jevons and the coal question. Organ Environ:93 et seq

CMEPSP (2009) Report of the commission on the measurement of economic performance and social progress. Paris

Costanza R (2009) A new development model for a "full" world. Development 52:369 et seq

Costanza R (2014) Commentary on monetizing nature. http://greattransition.org/commentary/robert-costanza-monetizing-nature-barbara-unmuessig

Daly H (1996) Steady-State economics: a new paradigm. New Lit Hist 24:811 et seq

Deane P (1978) The evolution of economic ideas. Cambridge University Press, Cambridge

Delhey J (2013) Vom BIP zum Glück – Wohlbefinden als neues gesellschaftspolitisches Ziel. In: Wegner G (ed) Wohlstand, Wachstum, Gutes Leben. Wege zu einer Transformation der Ökonomie. Metropolis-Verl, Marburg, p 147 et seq

Deutsche Bank Research (2009) Automobilindustrie am Beginn einer Zeitenwende. EU Monitor 62

Dietz A (2005) Homo oeconomicus: Theologische und wirtschaftsethische Perspektiven auf ein ökonomisches Modell. Gütersloher Verl.-Haus, Gütersloh

Doliwa K (2005) William of Occam and Thomas Hobbes – on the nature of general concepts, studies in logic. Gram Rhetoric:101 et seq

Drexl J (1998) Die wirtschaftliche Selbstbestimmung des Verbrauchers. Eine Studie zum Privat- und Wirtschaftsrecht unter Berücksichtigung gemeinschaftsrechtlicher Bezüge. Mohr Siebeck, Tübingen

Dyckhoff H, Ahn H (2006) Kosten-Nutzen-Analyse. In: Verlagsgruppe Handelsblatt (ed) Wirtschaftslexikon. Das Wissen der Betriebswirtschaftslehre. Schäffer-Poeschel, Stuttgart, p 3196 et seq

Easterlin R (1995) Will raising income of all increase the happiness of all? J Econ Behav Organ:35 et seq

Easterly W (2003) The political economy of growth without development: a case study of Pakistan. In: Rodrik D (ed) In search of prosperity – analytic narratives on economic growth. Princeton University Press, Princeton, p 439 et seq

Economist (2010) Daily chart – money and happiness. The Economist online, 25.09.2010. http://www.economist.com/blogs/dailychart/2010/11/daily_chart_1/print

Ekardt F, Wieding J (2016) Rechtlicher Aussagegehalt des Paris-Abkommens: Eine Analyse der einzelnen Artikel. Zeitschrift für Umweltpolitik Sonderheft:36 et seq

Fatheuer T (2014) Neue Ökonomie der Natur. Eine kritische Einführung. Heinrich-Böll-Stiftung, Berlin

Feld L, Fischer J, Kirchgässner G (2006) The effect of direct democracy on income redistribution – evidence for Switzerland, CESifo working paper No. 1837. Munich

Felderer B, Homburg S (2005) Makroökonomik und neue Makroökonomik, 9th edn. Springer, Berlin et al

Fronius M (2001) Die Ursachen des taiwanesischen Wirtschaftswunders. Eine systematische Betrachtung. Berlin

Gabler-Verlag (ed) (1997) Gablers Wirtschaftslexikon, 14th edn. Wiesbaden

Gausemeier J, Wiendahl H-P (eds) (2011) Wertschöpfung und Beschäftigung in Deutschland. Berlin

Glaser J (2014) Der Werturteilsstreit in der modernen Nationalökonomie. Max Weber, Werner Sombart und die Ideale der Sozialpolitik. Marburg

Gossen HH (1854) Entwicklung der Gesetze des menschlichen Verkehrs und der daraus fließenden Regeln für menschliches Handeln. Braunschweig

Grossmann M et al (2010) Ökonomische Bewertung naturverträglicher Hochwasservorsorge an der Elbe. Münster
Grote R, Sheikh A (2010) Deforestation and climate change. Congressional Research Service, Washington, DC
Haack S (2011) Verfassungshorizont und Taburaum. Archiv des öffentlichen Rechts 136:365 et seq
Hansjürgens B et al (2015) Inwertsetzung biologischer Vielfalt. Naturschutzanliegen in andere Politikbereiche integrieren. Metropolis-Verlag, Marburg
Hanusch H (1987) Nutzen-Kosten-Analyse. Munich
Hart HLA (1955) Are there any natural rights? Philos Rev:175 et seq
Hartard S, Stahmer C (eds) (2002) Sozio-ökonomische Berichtssysteme. Marburg
Heertje A, Wenzel H-D (2002) Grundlagen der Volkswirtschaftslehre, 6th edn. Berlin et al
Heidbrink L, Seele P (2011) Kommerzialisierung der Moral. Unter welchem Umständen sich moralische Mittel für Marktzwecke einsetzen lassen. In: Kettner M, Koslowski P (eds) Ökonomisierung und Kommerzialisierung der Gesellschaft. Wirtschaftsphilosophische Unterscheidungen. Wilhelm Fink, Munich, p 237 et seq
Hobbes T (1966) De Homine, Opera Philosophica, New edition. Aalen
Hoche H-U (2011) Die Universalisierte Goldene Regel als Grundlage einer kulturübergreifenden Moral und Moralerziehung. Jahrbuch für Recht und Ethik 19:471 et seq
Hoffmann J (1986) Die großen Wirtschaftsdenker. Düsseldorf
Horn K (2009) Der Realismus des Adam Smith. FAZ vom 16.01.2009, p 12
Hösle V (2014) Der Wert des eigenen Glücks. Inf Philos 4:8 et seq
Huber P (2001) Die Vorgaben des Grundgesetzes für kommunale Bürgerbegehren und. Bürgerentscheide. Archiv des öffentlichen Rechts 126:165 et seq
Hume D (1978) A treatise of human nature. hg. von L.A. Selby-Bigge, 2nd edn. London
Husted S, Melvin M (2009) International economics, 8th edn. Boston
Hüther M (2006) Adam Smith. In: Hüther M (ed) Klassiker der Ökonomie. Von Adam Smith bis Amartya Sen. Bonn, p 31 et seq
IMF (2010) Regional economic outlook – Europe moving to restore growth and sustainability. IMF European Department, 11.05.2010. http://www.imf.org/external/pubs/ft/survey/so/2010/CAR051010A.htm
Jaeger C (2011) Wachstum – wohin? Eine kurze Geschichte des 21. Jahrhunderts. Oekom-Verl, Munich
Jens U, Romahn H (eds) (2010) Methodenpluralismus in den Wirtschaftswissenschaften. Metropolis-Verl, Marburg
Jevons W (1965) The coal question, 3rd edn. A.M. Kelley, New York
Jones C (2002) Introduction to economic growth, 2nd edn. W.W. Norton, New York
Kahl W (2010) Grundrechte. In: Depenheuer O, Grabenwarter C (eds) Verfassungstheorie. Tübingen, p 809 et seq
Kahl W, Welke B (2010) Über die unveränderte Notwendigkeit einer integrierten Vorhabengenehmigung und deren Regelungsstandort. Deutsches Verwaltungsblatt 125:1414 et seq
Keynes JM (1923) A tract on monetary reform. London
Khazzoom D (1980) Economic implications of mandated efficiency in standards for household suppliances. Energy J:21 et seq
Kirchner C (2013) Wachstum, Ökonomik und Eigennutz. In: Wegner G (ed) Wohlstand, Wachstum, Gutes Leben. Wege zu einer Transformation der Ökonomie. Metropolis-Verl, Marburg, p 169 et seq
Klingholz R (2012) Sklaven des Wachstums. Frankfurt a.M. et al
Koesters P-H (1985) Ökonomen verändern die Welt. Munich
Kurz H (1996) Geiz der Natur. In: Piper N (ed) Die großen Ökonomen. Stuttgart, p 37 et seq
Kurz H (2008) David Ricardo. In: Kurz H (ed) Klassiker des ökonomischen Denkens, Bd. 1. Beck, Munich, p 120 et seq
Kuznets S (1966): Modern economic growth. New Haven

Linneweber V, Lantermann E-D, Kals E (eds) (2010) Spezifische Umwelten und umweltbezogenes Handeln. Hogrefe, Göttingen et al

Löhr D (2010) Nullwachstum und Nullzins – Renaissance einer alten Idee. Zeitschrift für Sozialökonomie 2:3 et seq

Lottermoser F (2014) Der reflexive Konsument. Gesellschaftsinteresse im 21. Jahrhundert. Nomos, Baden-Baden

Lübbe W (2002) Epistemische Pflichten in der „Wissensgesellschaft". In: Engel C, Halfmann J, Schulte M (eds) Wissen – Nichtwissen – Unsicheres Wissen. Nomos, Baden-Baden, p 145 et seq

Luterbacher-Maineri C (2008) Adam Smith – theologische Grundannahmen. Academic, Fribourg

MacAskill W (2016) Gutes besser tun. Wie wir mit effektivem Altruismus die Welt verändern können. Ullstein eBooks, Berlin

Maddison A (2006) The world economy. OECE Development Centre Studies, Paris

Mankiw G (2004) Grundzüge der Volkswirtschaftslehre, 3rd edn. Schäffer-Poeschel, Stuttgart

Martinez-Alier J (1987) Ecological economics. Oxford

Materialflows (2011a) Trends in global resource extraction, GDP and material intensity 1980–2007. http://www.materialflows.net/index.php?option=com_content&task=view&id=32&Itemid=48

Materialflows (2011b) Material intensity by world region 1980 vs. 2007. http://www.materialflows.net/index.php?option=com_content&task=view&id=78&Itemid=47

Materialflows (2011c) Global resource extraction per capita by world region 2007. http://www.materialflows.net/index.php?option=com_content&task=view&id=30&Itemid=46

Maus I (2011) Über Volkssouveränität. Elemente einer Demokratietheorie. Suhrkamp, Berlin

Mayert A (2013) Ist bewusste Wachstumsbegrenzung eine sinnvolle Strategie? In: Wegner G (ed) Wohlstand, Wachstum, Gutes Leben. Wege zu einer Transformation der Ökonomie. Metropolis-Verl, Marburg, p 89 et seq

Meadows D et al (1972) Die Grenzen des Wachstums. Bericht des Club of Rome zur Lage der Menschheit. Deutsche Verlags-Anstalt, Stuttgart

Möller A (2013) Das grüne Gewissen. Wenn die Natur zur Ersatzreligion wird. Hanser, Munich

Nell E, Semmler W, Rezai A (2011) Economic growth and climate change – cap-and-trade or emission tax? In: Altvater E, Brunnengräber A (eds) After cancun. Climate governance or climate conflicts. VS Verlag für Sozialwissenschaften, Wiesbaden, p 95 et seq

Neumann M (1994) Neoklassik. In: Issing O (ed) Geschichte der Nationalökonomie, 3rd edn. Munich, p 255 et seq

Newell P, Paterson M (2011) Climate capitalism. In: Altvater E, Brunnengräber A (eds) After cancun. Climate governance or climate conflicts. VS Verlag für Sozialwissenschaften, Wiesbaden, p 23 et seq

Occam W (1996) Summa Logicae I, 15, 10. In: Texte zur Theorie der Erkenntnis und der Wissenschaft. Munich

OFID (2009) Biofeuls and food security, OFID pamphlet series. Wien

Osterkamp T (2004) Juristische Gerechtigkeit. Rechtswissenschaft jenseits von Positivismus und Naturrecht. Mohr Siebeck, Tübingen

Petersen T, Faber M (2013) Karl Marx und die Philosophie der Wirtschaft. Bestandsaufnahme – Überprüfung – Neubewertung. Freiburg et al

Polimeni J et al (2009) The myth of resource efficiency. Routledge, London

Raphael DD (1991) Adam Smith. Frankfurt a.M

Raupach M et al (2007) Global and regional drivers of accelerating CO_2 emissions. Proc Natl Acad Sci USA 104:10288

Recktenwald HC (1989) Adam Smith. In: Starbatty J (ed) Klassiker des ökonomischen Denkens, Bd. 1. Beck, Munich, p 134 et seq

Regli S (1921) Die Staatsgewalt bei Thomas Hobbes im Lichte seines Nominalismus. Freiburg

Reuter N (2007) Wachstumseuphorie und Verteilungsrealität. Wirtschaftspolitische Leitbilder zwischen Gestern und Morgen. Metropolis, Marburg

Ricardo D (1994) Über die Grundsätze der politischen Ökonomie und der Besteuerung, New edition. Marburg

Ridoutt B et al (2015) Areas of concern. A new paradigm in life cycle assessment for the development of footprint metrics. Int J Life Cycle Assess 21:276
Rogall H (2014) 100-%-Versorgung mit erneuerbaren Energien. Bedingungen für eine globale, nationale und kommunale Umsetzung. Metropolis-Verlag, Marburg
Rolle R (2005) Homo oeconomicus. Wirtschaftsanthropologie in philosophischer Perspektive. Königshausen & Neumann, Würzburg
Roncaglia A, Sylos Labini P (2008) Geschichte des ökonomischen Denkens. Marburg
Ross IS (1998) Adam Smith – Leben und Werk. Verl. Wirtschaft und Finanzen, Düsseldorf
Ruckriegel K (2012) Glücksforschung – Konsequenzen für die (Wirtschafts-)Politik. Wirtschaftsdienst 92:129 et seq
Rusche T (1991) Philosophische versus ökonomische Imperative einer Unternehmensethik. Lit, Münster
Rusche T (2002) Aspekte einer dialogbezogenen Unternehmensethik. Dialogbezogene Begründung – christliche Motivation – verantwortungsvernünftige Praxis, 2nd edn. Lit, Münster
Sachverständigenrat für Umweltfragen (2012) Umweltgutachten 2012. Berlin
Samuelson P, Nordhaus W (1998) Volkswirtschaftslehre. Ueberreuter, Frankfurt a.M. et al
Sanden J (2009) Die New-Governance-Ansätze in der Europäischen Wirtschaftspolitik am Beispiel der Energiepolitik. Tectum, Marburg
Sassone P, Schaffer W (1978) Cost-benefit analysis: a handbook. Academic, New York
Saunders H (1992) The Khazzoom-Brookes postulate and neoclassical growth. Energy J 13:131 et seq
Schäfer F et al (2006) Ecological footprint and biocapacity. European Commission, Luxemburg
Schewe G, Nienaber A-M, Buschmann A, Liesenkötter B (2012) Alles nur Greenwashing? Wie glaubwürdig berichten Unternehmen über ihr Nachhaltigkeitsengagement? Zeitschrift für Umweltpolitik und Umweltrecht 35:1 et seq
Schmidt K-H (1994) Merkantilismus, Kameralismus, Physiokratie. In: Issing O (ed) Geschichte der Nationalökonomie, 3rd edn. Munich, p 37 et seq
Schmidt T, Tarkian T (eds) (2011) Naturalismus in der Ethik. Perspektiven und Grenzen. Mentis, Paderborn
Schmidt-Bleek F (2014) Grüne Lügen. Nichts für die Umwelt, alles fürs Geschäft – wie Politik und Wirtschaft die Welt zugrunde richten. Munich
Schönwälder-Kuntze T (2011) Auf wen oder was antwortet Verantwortung? Zur Genealogie (und Pathologie) des Verantwortungsdenkens. Jahrbuch für Recht und Ethik 19:367 et seq
Schumann J, Meyer U, Wolfgang S (2007) Grundzüge der mikroökonomischen Theorie, 8th edn. Springer, Berlin et al
Schwalbe U (2008) Léon Walras. In: Kurz H (ed) Klassiker des ökonomischen Denkens, Bd. 1. Beck, Munich, p 242 et seq
Shue H (2014) Climate justice. Vulnerability and protection. Cambridge, MA
Sohn-Rethel A (2008) Das Ideal des Kaputten. Frickingen
Söllner F (2001) Die Geschichte des ökonomischen Denkens, 2nd edn. Springer, Berlin
Sommer M (2013) Wohlstand ohne Wachstum – eine sinnvolle Alternative? In: Wegner G (ed) Wohlstand, Wachstum, Gutes Leben. Wege zu einer Transformation der Ökonomie. Metropolis-Verl, Marburg, p 25 et seq
Stadermann H-J, Steiger O (2006) Allgemeine Theorie der Wirtschaft, Band 1: Schulökonomik. Tübingen
Stiglitz J (2009) GDP Fetishism, Project Syndicate. http://www.project-syndicate.org/commentary/stiglitz116/English (15.02.2017)
Straubhaar T (2006) Wohlstand durch Freiheit. In: Hüther M (ed) Klassiker der Ökonomie. Von Adam Smith bis Amartya Sen. Bonn, p 46 et seq
Streminger G (1989) Adam Smith. Hamburg
Thaler R, Sunstein C (2008) Nudge. Improving decisions about health, wealth, and happiness. New Haven
Trepl L (2012) Die Idee der Landschaft. Eine Kulturgeschichte von der Aufklärung bis zur Ökologiebewegung. Bielefeld

Tschentscher A (2003) Der Konsensbegriff in Vertrags- und Diskurstheorien. In: van Aaken A, Schmidt-Lübbert S (eds) Beiträge zur ökonomischen Theorie im Öffentlichen Recht. Deutscher Universitätsverlag, Wiesbaden, p 119 et seq

Turner K (2007) Limits to CBA in UK and European Environmental Policy – retrospect and future prospects. Environ Resour Econ 37:253 et seq

van Suntum U (2005) Die unsichtbare Hand. Ökonomisches Denken gestern und heute, 3rd edn. Berlin et al

van Suntum U (2008) William Stanley Jevons. In: Kurz H (ed) Klassiker des ökonomischen Denkens, Bd. 1, Munich, p 267 et seq

von Braun J (2015) Welternährung und Nachhaltigkeit. Herausforderungen und Strategien für das 21. Jahrhundert. Oekom Verlag, Munich

von Wright GH (2008) Erklären und Verstehen. Europ. Verl.-Anst, Hamburg

Wackernagel M (2009) Ecological creditors and debitors. Environ Poverty Times:1 et seq

Weber M (1973) Wirtschaft und Gesellschaft, 5th edn. Tübingen

Wegner G (2013) Zur Problematik des Wachstums. In: Wegner G (ed) Wohlstand, Wachstum, Gutes Leben. Wege zu einer Transformation der Ökonomie. Metropolis, Marburg, p 7 et seq

Weil D (2005) Economic growth. Munich

Welfens P (2008) Grundlagen der Wirtschaftspolitik: Institutionen – Makroökonomik – Politikkonzepte, 3rd edn. Springer, Berlin et al

Wieland J (2011) Die Entdeckung der Ökonomie. Kategorien, Gegenstandsbereiche und Rationalitätstypen, 2nd edn. Metropolis, Marburg

Wohltmann H-W (2007) Grundzüge der makroökonomischen Theorie, 5th edn. Oldenbourg, Munich

Index

F. Ekardt, *Economic Evaluation, Cost-Benefit Analysis, Economic Ethics*, Environmental Humanities: Transformation, Governance, Ethics, Law 2,
https://doi.org/10.1007/978-3-030-99284-2

MIX
Papier aus verantwortungsvollen Quellen
Paper from responsible sources
FSC® C105338

If you have any concerns about our products,
you can contact us on
ProductSafety@springernature.com

In case Publisher is established outside the EU,
the EU authorized representative is:
Springer Nature Customer Service Center GmbH
Europaplatz 3, 69115 Heidelberg, Germany

Printed by Libri Plureos GmbH
in Hamburg, Germany